EXPLORING THE WAY LIF

The Introduction and Patterns in the **CONCEPTS** section have their own web elaboration found online at the **WebConnections** site (www.jbpub.com/connections). For the Introduction and for every numbered Pattern you'll find corresponding web links. Jones and Bartlett Publishers monitors these links weekly to ensure that they are current and open.

On the ***WebConnections*** *home page, choose the Pattern you wish to explore further by clicking on its number and name.*

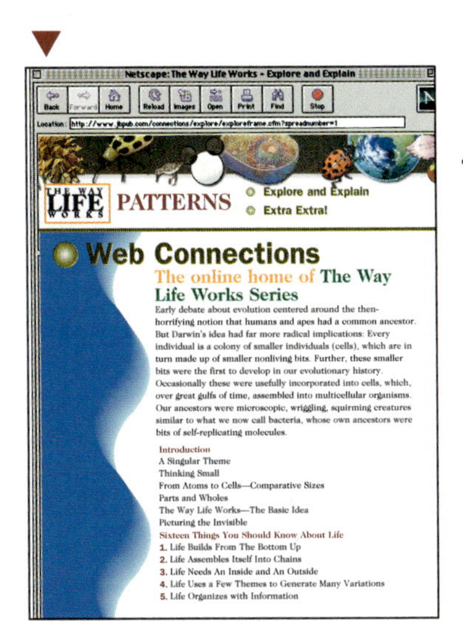

For each Pattern, you will find two different kinds of WebConnections.

- *Explore and Explain*
- *Extra! Extra!*

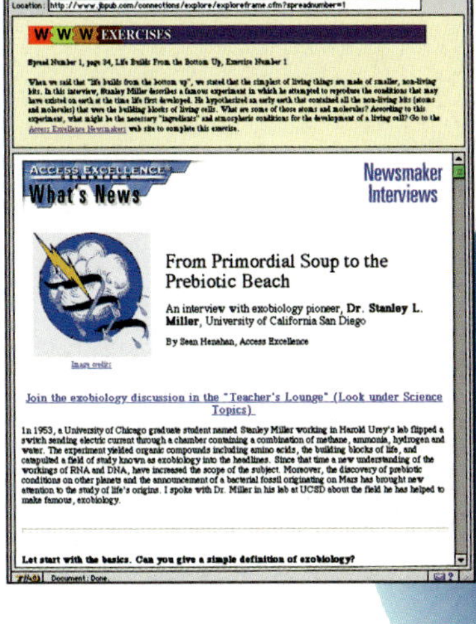

▲ **Explore and Explain** asks questions about **CONCEPTS** material, and hyperlinks you directly to sites where you can get expanded information and find the answers to these questions.

Whenever you see this WebConnections icon, you know that there is material to be explored on the Way Life Works' website. Just enter the URL **www.jbpub.com/connections** on a World Wide Web browser such Internet Explorer to get to the book's home page. Save the URL in your "favorites" file.

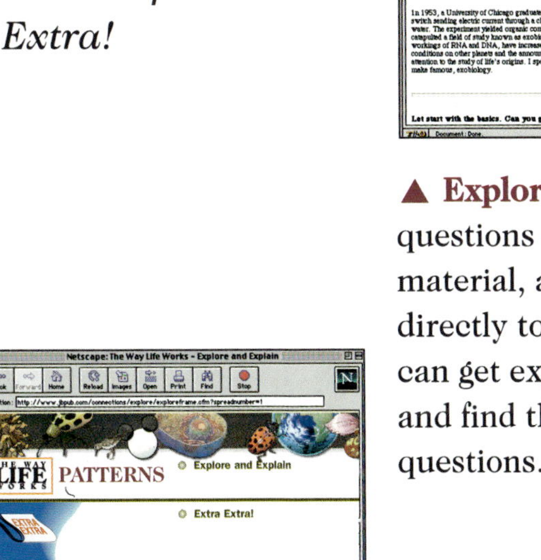

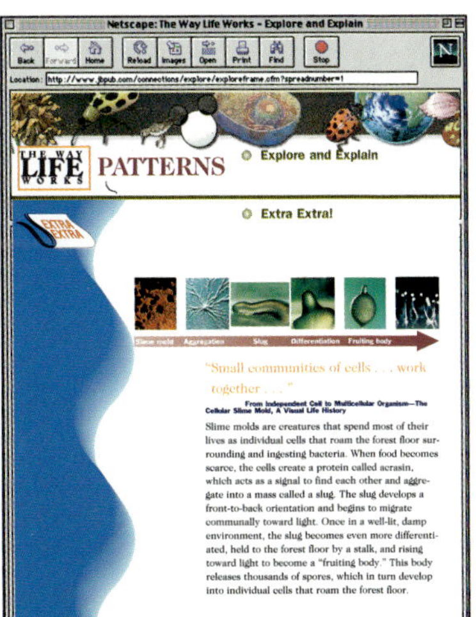

▲ **Extra! Extra!** brings you new discoveries or recent research that reveals scientific curiosity at work. Here you will also find illustrations and photographs that provide further visual details of topics discussed in your chosen Pattern.

World Headquarters
Jones and Bartlett Publishers
40 Tall Pine Drive
Sudbury, MA 01776
978-443-5000
info@jbpub.com
www.jbpub.com

Jones and Bartlett Publishers Canada
2100 Bloor St. West, Suite 6-272
Toronto, ON M6S 5A5
CANADA

Jones and Bartlett Publishers International
Barb House, Barb Mews
London W6 7PA
UK

PRODUCTION CREDITS
V.P., MANAGING EDITOR/DEVELOPMENTAL EDITOR Judith Hauck
V.P., COLLEGE EDITORIAL DIRECTOR Brian McKean
DIRECTOR OF DESIGN AND PRODUCTION Anne Spencer
TEXT AND COVER DESIGNER Anne Spencer
DESIGN/PRODUCTION Rebecca Marks
DESIGN/PRODUCTION Stephanie Torta
WWW DESIGN/PRODUCTION Mark Rodrigues
V.P., DIRECTOR OF INTERACTIVE TECHNOLOGY Mike Campbell
ASSISTANT EDITOR Karen McClure

CONCEPTS pages reprinted by arrangement with Times Books, a division of Random House, Inc.

Library of Congress Cataloging-In-Publication Data
Hoagland, Mahlon B.
Patterns: sixteen things you should know about life / [Mahlon Hoagland, Bert Dodson].
p. cm. -- (The way life works series : v. 1)
ISBN 0-7637-0648-5
1. Life (Biology) I. Dodson, Bert. II. Title. III. Series.
QH501.H56 1999
570--dc21 99-10891
CIP

PHOTO CREDITS
p. 35 By permission of the President and Council of the Royal Society; p. 37, Dr. Tony Brain and David Parker/Science Photo Library/Photo Researchers ; p. 41, S. Dalton/Animals Animals; p. 45, Patrick Grace/Science Source/Photo Researchers; p. 46, Used with permission from Art Siegel, University of Pennsylvania.; p. 38, Rosalind Franklin/Cold Spring Harbor Laboratory Archives; p. 37, used with permission of Jones and Bartlett Publishers from Electron Microscopy, 2E by Bozzola/Russell; pp. 35, 36, used with permission of Jones and Bartlett Publishers from Electron Microscopy, 2E by Bozzola/Russell; p. 36, used with permission of Jones and Bartlett Publishers from Electron Microscopy, 2E by Bozzola/Russell; p. 36, Molecules R us; p. 40, SUNY Nassua; p. 40, SUNY Nassau; p. 44, NASA; p. 43, Used with permission from Fox and Dose, Molecular and the Origin of Life, Freeman, San Francisco; p. 43, Used with permission from Fox and Dose, Molecular and the Origin of Life, Freeman, San Francisco; p. 45, (c)Kim Taylor/Bruce Coleman/PNI; p. 39, John D. Cunningham/Visuals Unlimited; p. 41, Photodisc; p. 36, Science Photo Library/Professor ; p. Stanley Cohen/Science Photo Library/; Photo Researchers; p. 39, Lee D. Simon/Science Photo Library/Photo Researchers; p. 38, Centre National de Recherches Iconographiques; p. 45, Ralph Reinhold/Animals Animals/Earth Scenes; p. 44, US Geological Survey; p. 40, Marc Rampulla/Peter Arnold; p. 50, © Eric Sander 1997; p. 50, Used with permission from John Doebley, Plimoth Plantation Agriculture.; p. 59, David M. Phillips/Visuals Unlimited; p. 37, used with permission of Jones and Bartlett Publishers from Electron Microscopy, 2E by Bozzola/Russell; p. 51, Stanley Flegler/Visuals Unlimited; p. 51, Stanley Flegler/Visuals Unlimited; p. 52, Used with permission from Rockefeller University Press.; p. 52, Used with permission from W. H. Freeman and Company.; p. 55, Used with permission from George Bernard, Animals Animals.; p. 54, used with permission of Jones and Bartlett Publishers from Electron Microscopy, 2E by Bozzola/Russell; p. 55, NASA; p. 57, Charles Palek; p. 50, © Eric Sander ; p. 56, Jim Mauseth; p. 56, Butch Gemin; p. 58, David M. Phillips/Visuals Unlimited; p. 60, Todd Barkman, University of Texas; p. 61, J. Frederick Grassle/Woods Hole Oceanographic Insitution; p. 62, NASA/Galileo Imaging Team; p. 62, IMP Team/NASA/JPL;

Printed in the United States of America
03 02 01 00 99 10 9 8 7 6 5 4 3 2 1

PATTERNS

Sixteen Things You Should Know About Life

MAHLON HOAGLAND
BERT DODSON

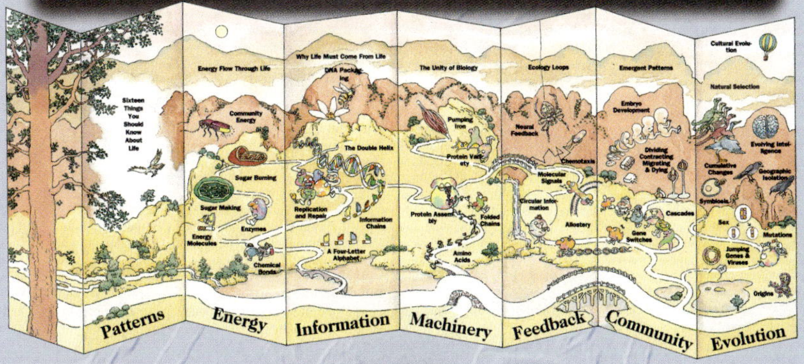

JONES AND BARTLETT PUBLISHERS
Sudbury, Massachusetts
BOSTON TORONTO LONDON SINGAPORE

Not long after the publication of **The Way Life Works** By Times Books, Jones and Bartlett conceived the idea of expanding and restructuring the book as a series of textbooks for college non-science majors and high school juniors and seniors. To our delight they have followed through with singular verve and originality. We have worked closely with them to maintain the spirit of our original effort as new material was added to satisfy the needs of students and faculty.

We tried to convey that spirit in an Author's Note for **The Way Life Works**, which we quote in full here.

Authors' Note

When we—biologist and artist—first met in 1988, we discovered that we shared a fascination with the unity of life—how, deep down, all living creatures, from bacteria to humans, use the same materials and ways of doing things.

We began exploring ways we might share our wonder with others, and came to believe we could achieve our purpose through an intimate merging of science and art. In the process, we hoped to persuade our audience that a deeper understanding of nature would enhance their appreciation of its beauty—and thereby enrich their lives.

Scientist as teacher and artist as student explained, questioned, searched, and argued. One day, Bert emerged with a two-page spread of pictures and Mahlon got a new vision of what he thought he knew; artist became teacher, scientist became student. Our confidence grew. We sifted, sorted, and pieced together our interpretation of the way life works.

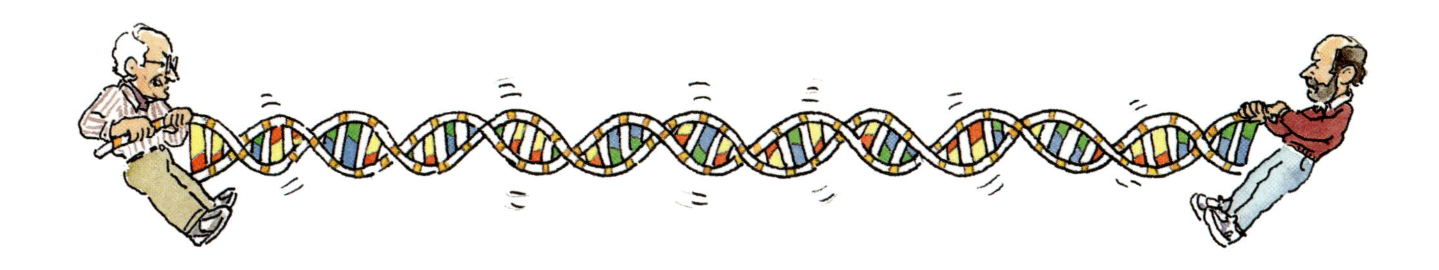

The scientist wants to leave the reader with a feeling of awe and pride in the achievements of scientific exploration, in the human potential for ever deeper understanding. The artist, on the other hand, sees the possibility that an appreciation for our oneness with the living world can guide our individual actions as we shape our collective future. We hope our readers will be moved by both.

Acknowledgments

The academic edition of **The Way Life Works Series** has clearly been a labor of love for a team of innovative and creative people at Jones and Bartlett.

Brian McKean enthusiastically accepted the challenge of publishing a very different way of teaching the core concepts of biology, contributing imaginative ideas for presenting the content. Anne Spencer designed the cover and Connections pages, and did all the page layout. Rebecca Marks proofread and did photo research, and Mark Rodrigues designed the Web pages. We are impressed at how beautifully they integrated new material into this edition of the book.

We want to express our sincere thanks to Judy Hauck. Before she joined Jones and Bartlett in 1997 she had been deeply interested in, and helpful to us, in developing **The Way Life Works Series** and we are grateful that her continuing involvement has been largely responsible for bringing the present project to fruition—one of us (Mahlon Hoagland) is proud that Judy is a first-rate editor and his daughter.

The capacity to tolerate complexity and welcome contradiction, not the need for simplicity and certainty, is the attribute of an explorer. Centuries ago, when some people suspended their search for absolute truth and began instead to ask how things worked, modern science was born. Curiously, it was by abandoning the search for absolute truth that science began to make progress, opening the material universe to human exploration. It was only by being provisional and open to change, even radical change, that scientific knowledge began to evolve. And ironically, its vulnerability to change is the source of its strength.

—*Heinz R. Pagels in* Perfect Symmetry: The Search for the Beginning of Time

. . . And it is a strange thing that most of the feeling we call religious, most of the mystical outcrying which is one of the most prized and used and desired reactions of our species, is really the understanding and the attempt to say that man is related to the whole thing, related inextricably to all reality, known and knowable. This is a simple thing to say, but a profound feeling of it made a Jesus, a St. Augustine, a Roger Bacon, a Charles Darwin, an Einstein. Each of them in his own tempo and with his own voice discovered and reaffirmed with astonishment the knowledge that all things are one thing and that one thing is all things—a plankton, a shimmering phosphorescence on the sea and the spinning planets and an expanding universe, all bound together by the elastic string of time.

—*John Steinbeck in* Log from the Sea of Cortez

PATTERNS
The Way Life Works

C O N T E N T S

CONCEPTS

CONNECTIONS

INTRODUCTION

Imagine you are walking along a deserted beach and you come upon the carcass of a whale. Time and tide and carrion birds have taken much of the flesh. Your first reaction might be a compassionate recognition of kinship. You might be curious about what happened — what was this whale's story?

As you examine the skeleton, a pattern strikes you. In each of the whale's front fins, the bones are arranged in three sections, with one bone in the section closest to the body, two parallel bones in the middle section, and five radiating branches of smaller bones in a more complex outer section. In fact, the bones of a whale's fin look very much like those of a human arm and hand. The proportions differ, but the pattern is remarkably similar.

How is it that a whale has arms like yours? And why does a whale have finger bones when it doesn't have fingers? Does this mean we're related to whales? Could it be that this limb pattern has been around longer than either whales…or humans?

A SINGULAR THEME

When we muse about life, what impresses us is its diversity — the sheer variety of organisms everywhere we look. Television programs and books about nature tend to celebrate the astonishing multiplicity of ways that life has adapted to our planet. This book's theme is different: It celebrates unity. It focuses on the things common to all forms of life, everywhere on earth.

Those homologous, or common, patterns in the bones of the human arm and the whale fin and, for that matter, in the bones of a bird's wing and a bat's wing, and even in the fossil remains of creatures that lived millions of years ago — are the first visible signs of unity. And the deeper we explore, the more signs we discover.

Every living being is either a cell or is made of cells: tiny, animate entities that gather fuel and building materials, produce usable energy, and grow and duplicate.

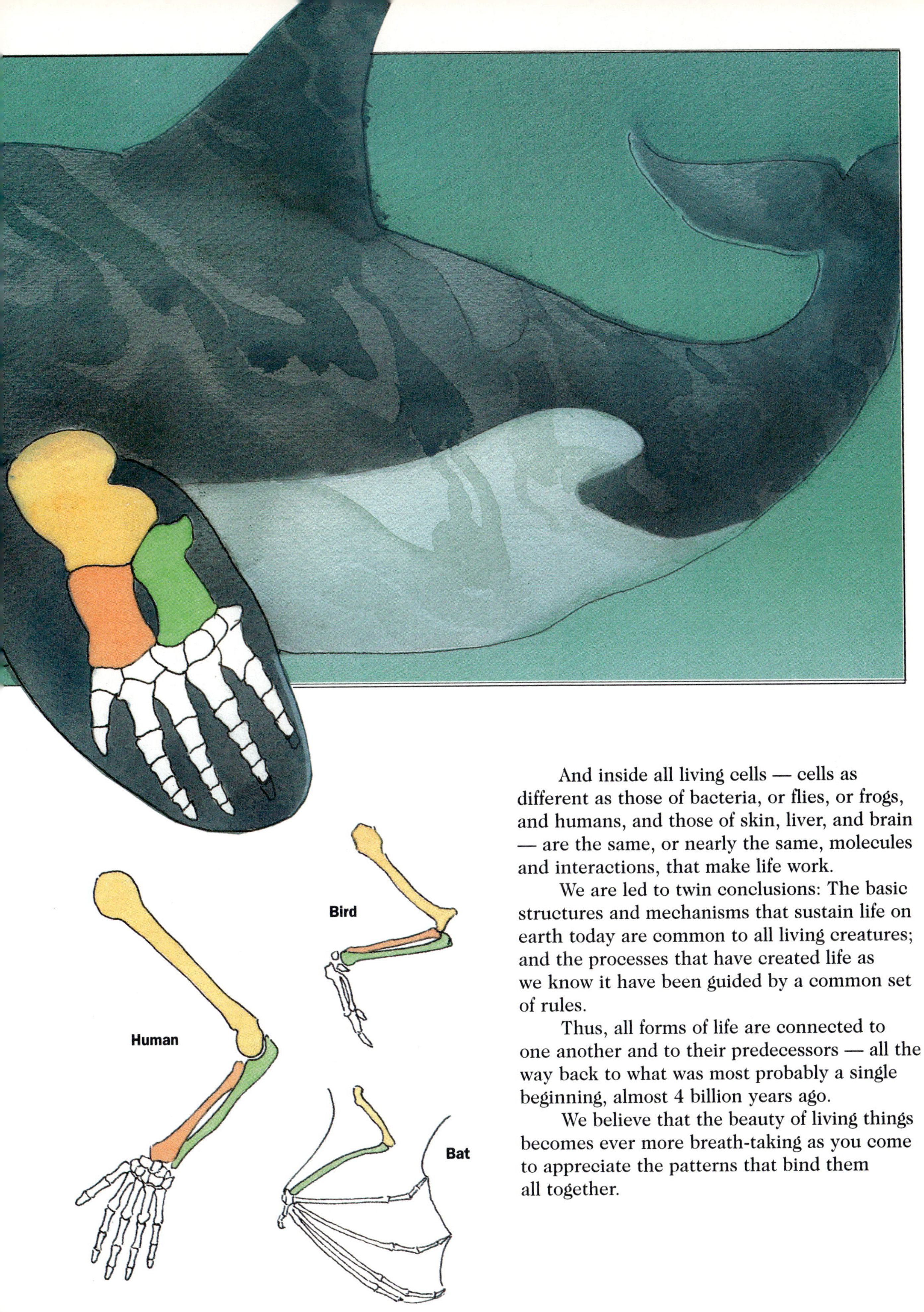

And inside all living cells — cells as different as those of bacteria, or flies, or frogs, and humans, and those of skin, liver, and brain — are the same, or nearly the same, molecules and interactions, that make life work.

We are led to twin conclusions: The basic structures and mechanisms that sustain life on earth today are common to all living creatures; and the processes that have created life as we know it have been guided by a common set of rules.

Thus, all forms of life are connected to one another and to their predecessors — all the way back to what was most probably a single beginning, almost 4 billion years ago.

We believe that the beauty of living things becomes ever more breath-taking as you come to appreciate the patterns that bind them all together.

THINKING SMALL

Much of this book takes place inside the cell. If you are unfamiliar with this microscopic landscape, understanding just how small and how numerous molecules are requires a considerable stretch of the imagination.

The great Scottish mathematician and physicist Lord Kelvin said: "Suppose that you could mark the molecules in a glass of water; then pour the contents of the glass into the ocean and stir the latter thoroughly so as to distribute the marked molecules uniformly throughout the seven seas; if then you took a glass of water anywhere out of the ocean, you would find in it about a hundred of your marked molecules."

Size and speed are related. Generally, the smaller an object is, the faster it can move. Water molecules, and all the other thousand or so kinds of molecules you have within you, swim about at stupendous speeds, flashing past each other and bumping into each other every millionth of a millionth of a second.

Life depends upon these frequent and vigorous collisions. It becomes a little easier to grasp the speed of the life-sustaining chemical transformations that constantly occur inside your cells (at the rate of thousands of events per second) when you realize that the participants move and collide millions of times faster.

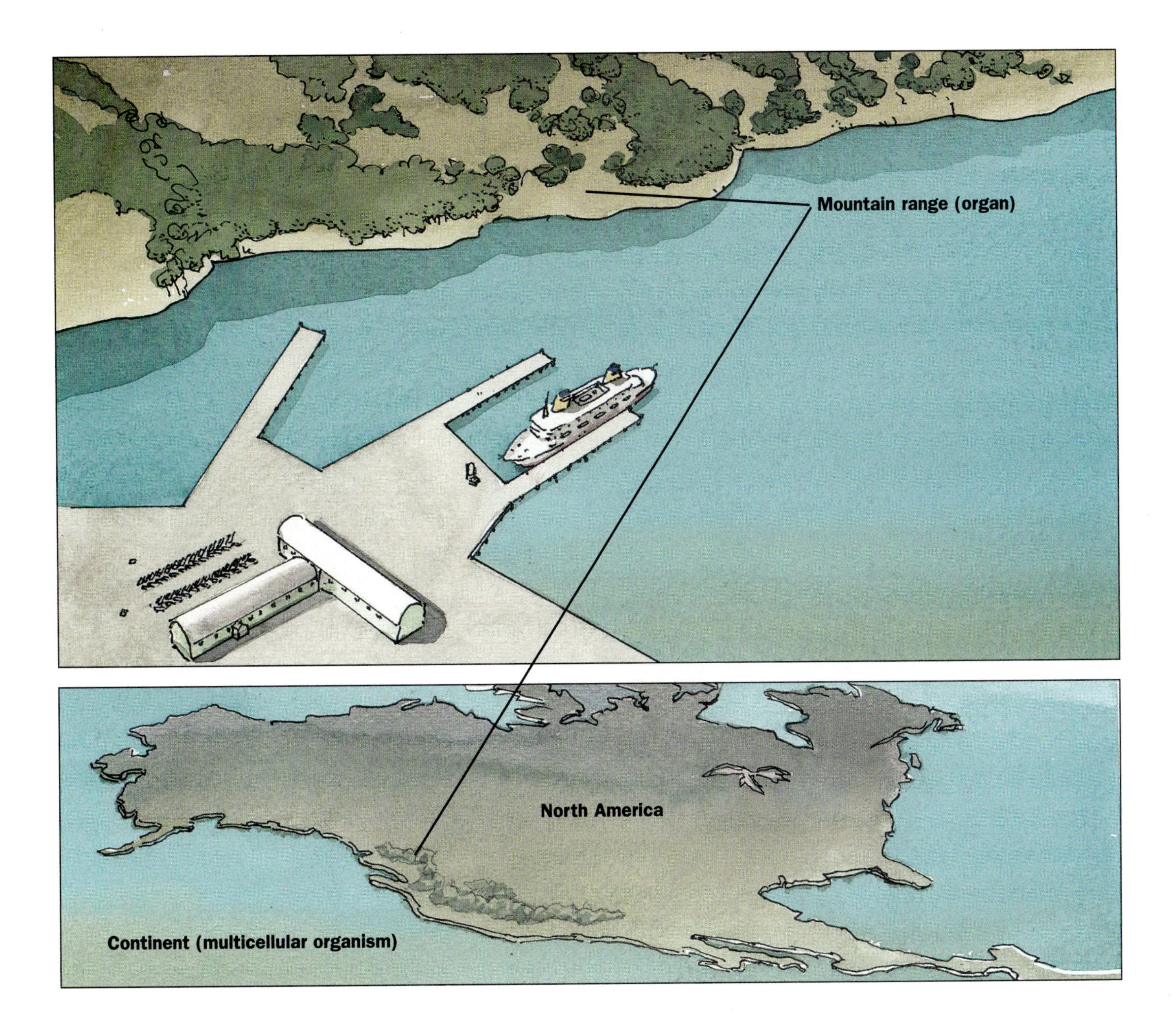

When we think about parts of the body, we tend to think of muscles, heart, brain, etc. The next step down in size brings us to the cells of which those parts are made. That drop in size is immense. Human cells are about ten times smaller than the point of a pin, and your body is composed of 5 trillion of them. And within each cell are multitudes of atoms, molecules, and structures made of molecules — the principal characters in our story.

As we introduce them, the picture above might help you to grasp their relative sizes.

Imagine you are standing on a pier. In one hand, you hold a BB — its size will represent an atom. In the other hand, you hold a marble — analogous to a simple molecule. Next to you is a cat — a chain molecule. Parked nearby is a tractor-trailer truck — a molecular structure. And tied up at the pier is an ocean liner — a cell. The pier is on the coast of North America — the whole continent being analogous in size to a human being.

On the following four pages we present a visual guide for distinguishing small things. Notice that four separate scales are necessary for spanning the range of size from atom to cell (a 200,000-fold jump).

CONNECTIONS»
page 35

From Atoms to Cells — Comparative Sizes

Scale 1. Atoms and Molecules

Magnified 50 Million Times

Atoms are the elemental units of which everything in the universe, living and non-living, is made. Atomic diameters range from one to a few hundred millionths of an inch.

Molecules are atoms bonded together. Much of life depends on three tiny molecules that have 2 to 3 atoms apiece: carbon dioxide (CO_2), the ultimate source of life's carbon atoms; oxygen (O_2), the gas crucial to energy generation in most life forms; and water (H_2O), the sea inside our cells in which life's machinery is bathed, and which aids chemical events inside our cells.

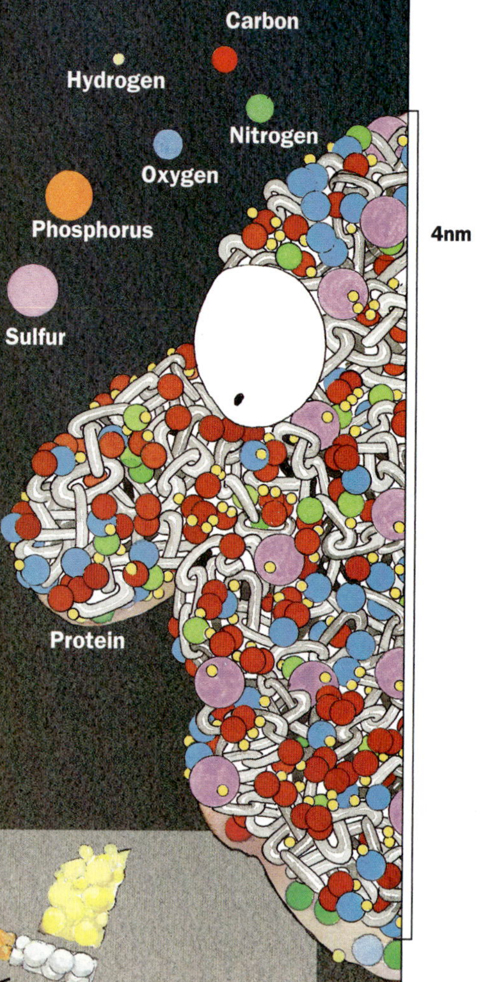

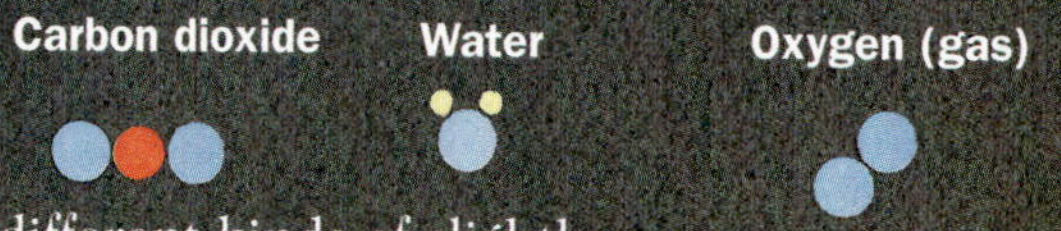

Carbon
Hydrogen
Nitrogen
Oxygen
Phosphorus
Sulfur
4nm
Protein

Carbon dioxide
Water
Oxygen (gas)

Roughly one thousand different kinds of slightly larger molecules made of 10 to 35 atoms are also found inside cells. These small molecules are either food (fuel) or building materials, or molecules that have been or will be food or building materials. We call all these simple molecules. The important ones in this book are sugars, nucleotides and amino acids.

Sugar
Nucleotide
Amino acid

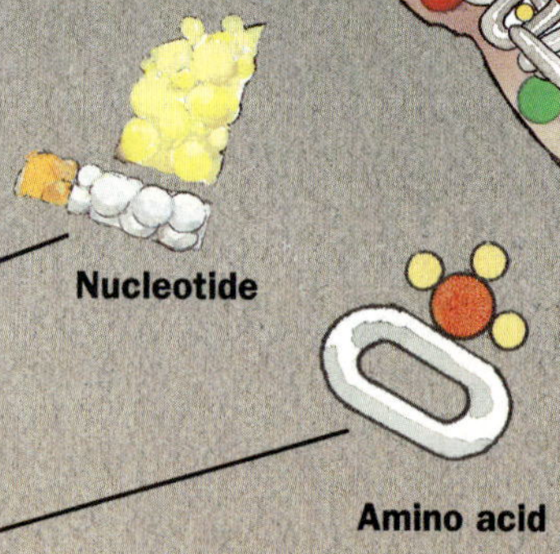

Nucleotide
Amino acid

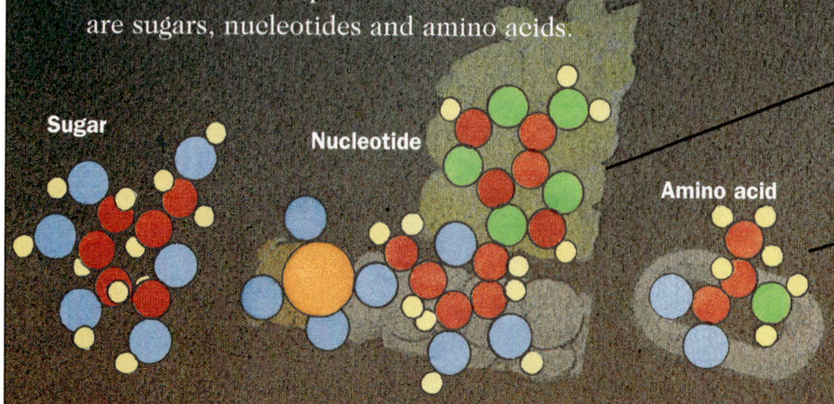

Throughout **Concepts** we depict nucleotides and amino acids as shown above. This best illustrates their function.

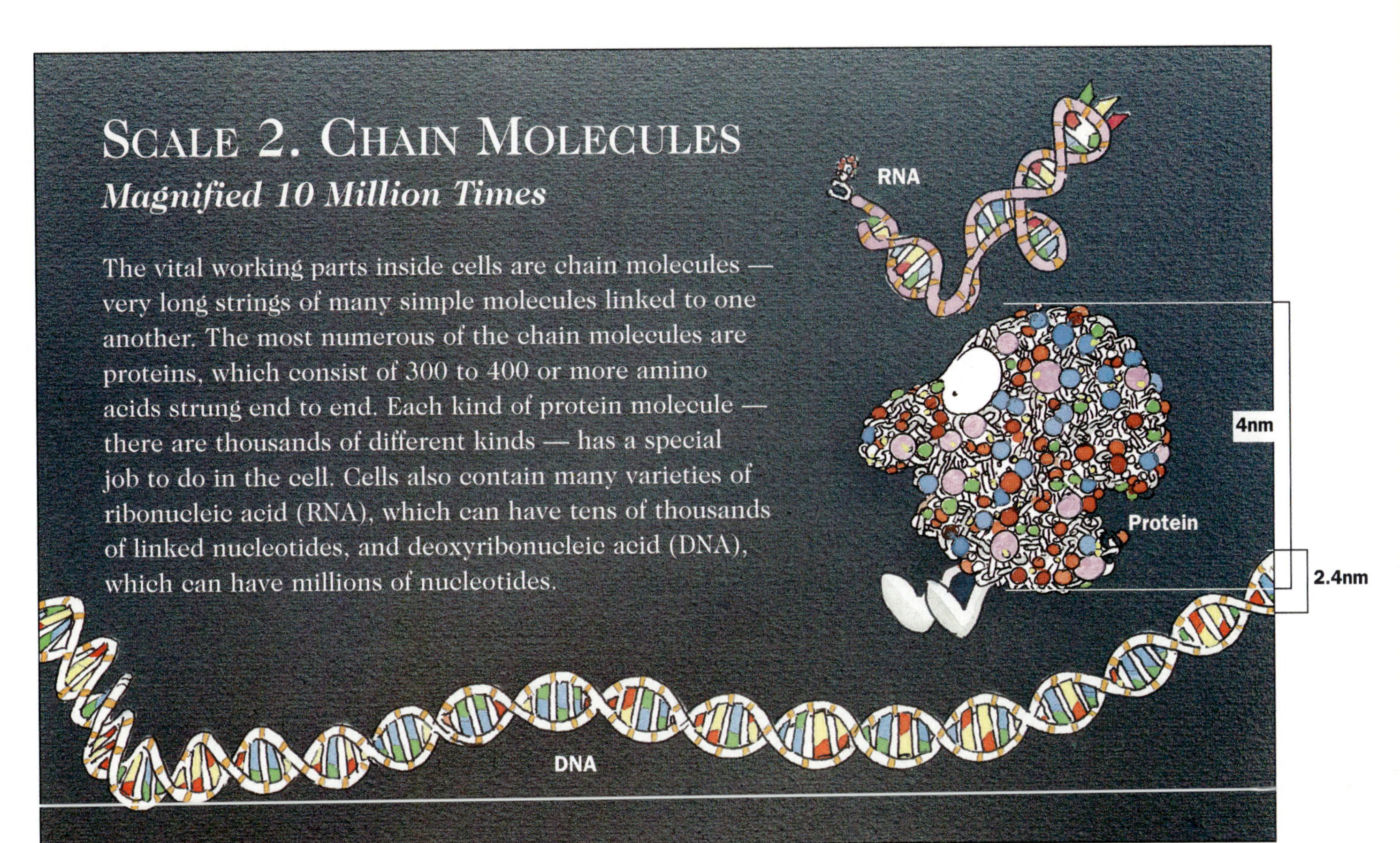

Scale 2. Chain Molecules

Magnified 10 Million Times

The vital working parts inside cells are chain molecules — very long strings of many simple molecules linked to one another. The most numerous of the chain molecules are proteins, which consist of 300 to 400 or more amino acids strung end to end. Each kind of protein molecule — there are thousands of different kinds — has a special job to do in the cell. Cells also contain many varieties of ribonucleic acid (RNA), which can have tens of thousands of linked nucleotides, and deoxyribonucleic acid (DNA), which can have millions of nucleotides.

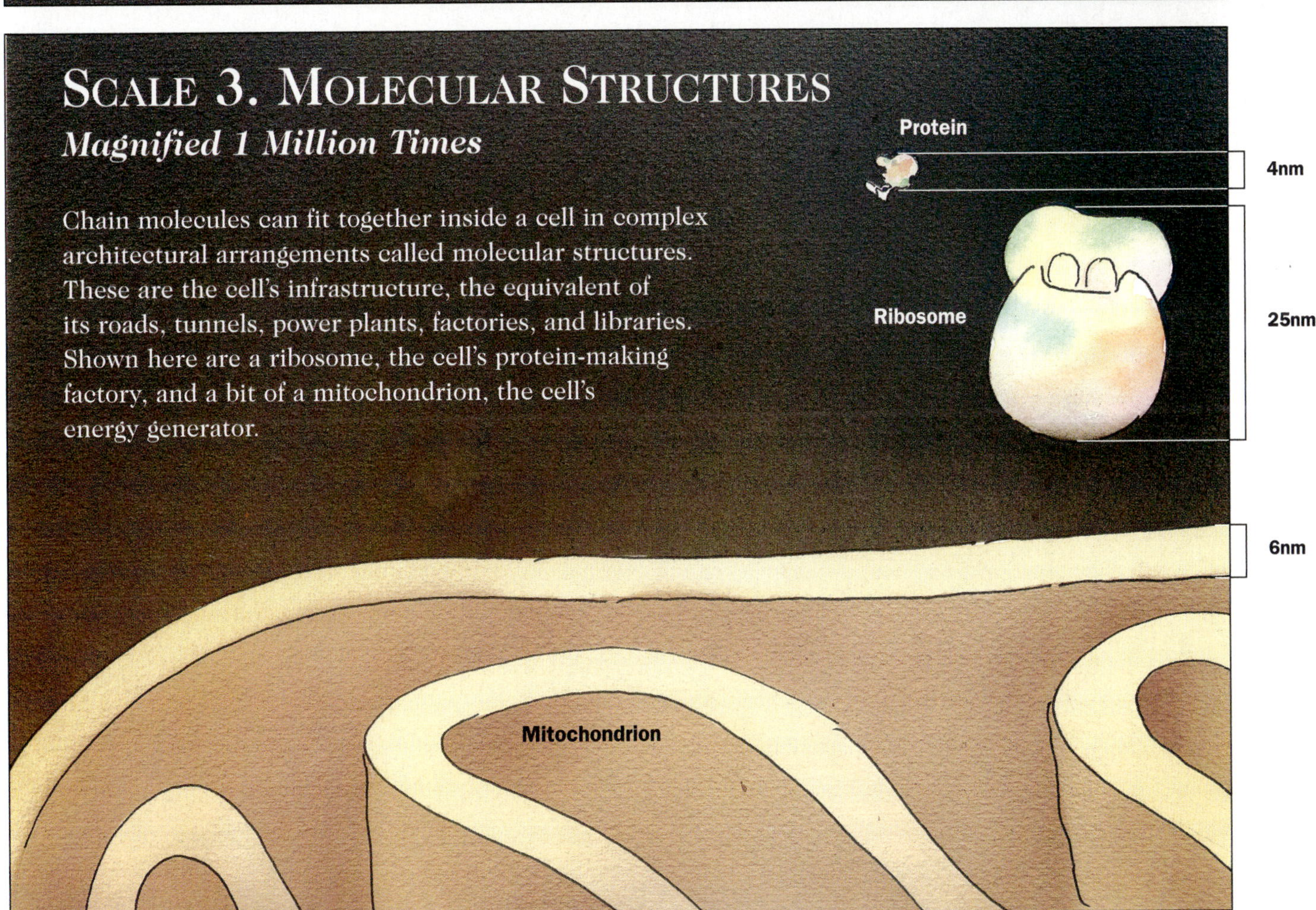

Scale 3. Molecular Structures

Magnified 1 Million Times

Chain molecules can fit together inside a cell in complex architectural arrangements called molecular structures. These are the cell's infrastructure, the equivalent of its roads, tunnels, power plants, factories, and libraries. Shown here are a ribosome, the cell's protein-making factory, and a bit of a mitochondrion, the cell's energy generator.

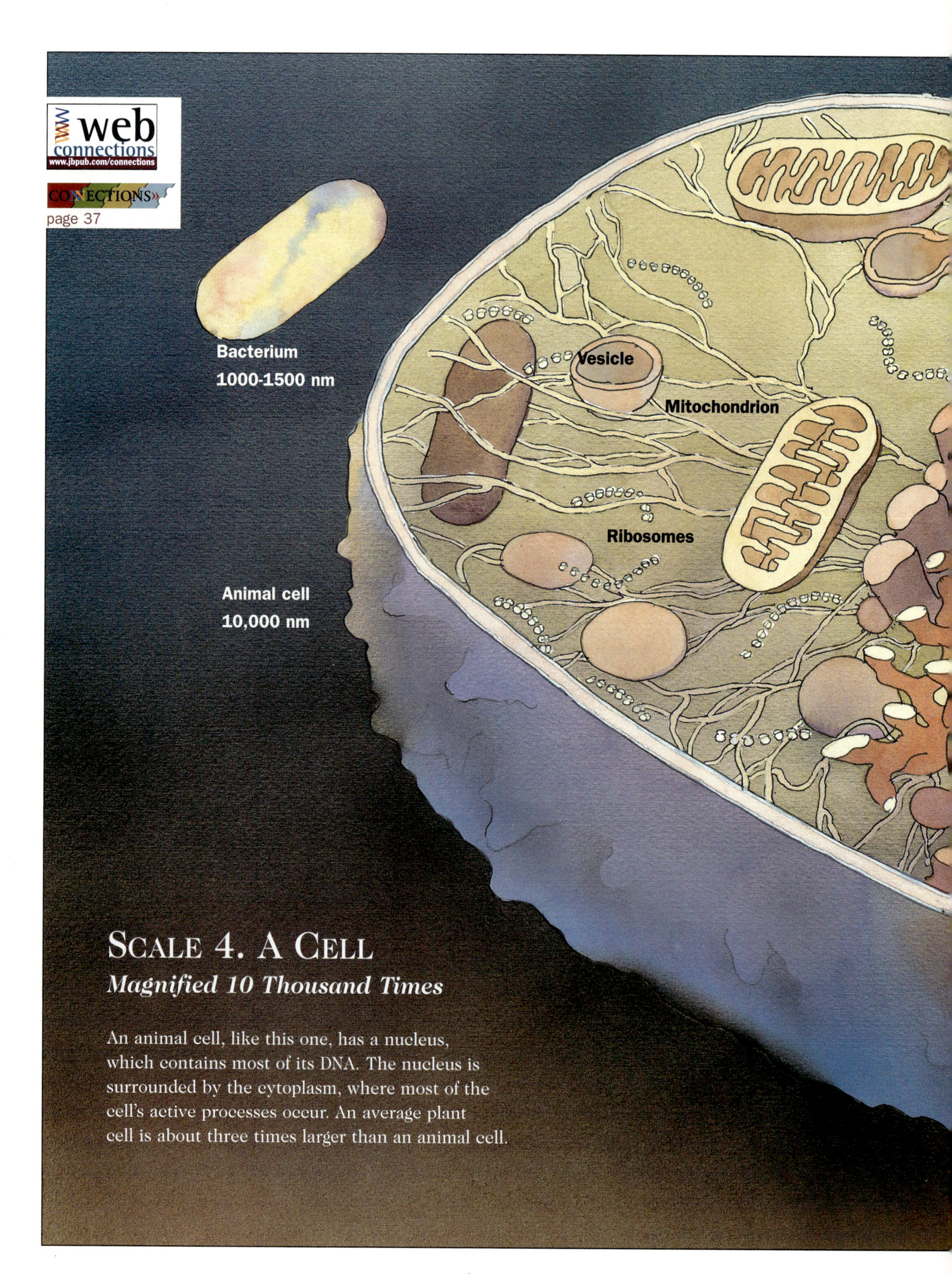

Scale 4. A Cell

Magnified 10 Thousand Times

An animal cell, like this one, has a nucleus, which contains most of its DNA. The nucleus is surrounded by the cytoplasm, where most of the cell's active processes occur. An average plant cell is about three times larger than an animal cell.

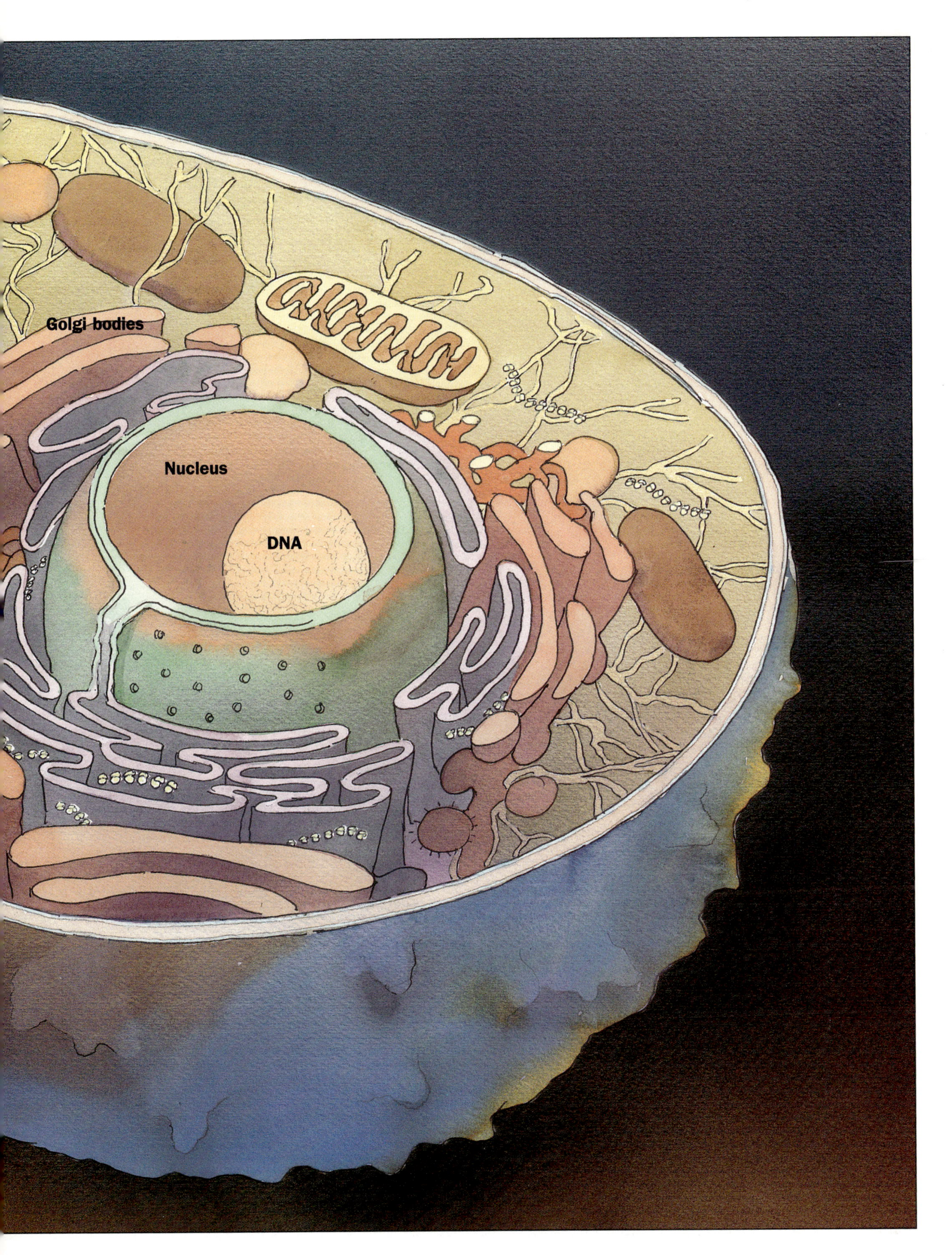
Golgi bodies
Nucleus
DNA

Parts and Wholes

It's useful to think of life's organization in levels, from the simple to the complex: atoms, simple molecules, chain molecules, molecular structures, cells — and onward and upward to organs, organisms, and communities of organisms. A higher level includes everything in the levels below it, as shown by the Russian dolls above.

Scientists find that knowing a lot about a lower level produces useful explanations of what's happening at the next higher level. To understand how your car works, you must know something about cylinders and spark plugs and fuel injection and how they interact.

This way of getting to understand the whole by learning about its parts, called reductionism, has produced in the last several decades an explosion of knowledge about what genes are and how they work, and how living processes are energized, informed, operated, and controlled. They are the "what" and "how" questions we take up in this book and it's succeeding volumes.

When we ask why things are the way they are we need to see things from the outside, and in relationship to others and to the surroundings. For example, why do birds have different beaks? To discover the answer, we need to study not just the birds themselves but, among other things, the food they eat. "Why" questions address patterns of connection in both space and time. They relate particularly to evolution — a subject touched on throughout *Patterns,* and explained in depth in Volume 7, *Evolution*.

Biochemists and molecular biologists tend to see themselves as reductionists, while naturalists and ecologists tend to take a holistic view. But, in fact, every scientist must shift his or her gaze regularly from the parts to the whole — from the trees to the forest — and back again.

We recommend that you try to be similarly fluid so that you can move back and forth with us as we shift from the tiny micro world to the larger macro world and back again.

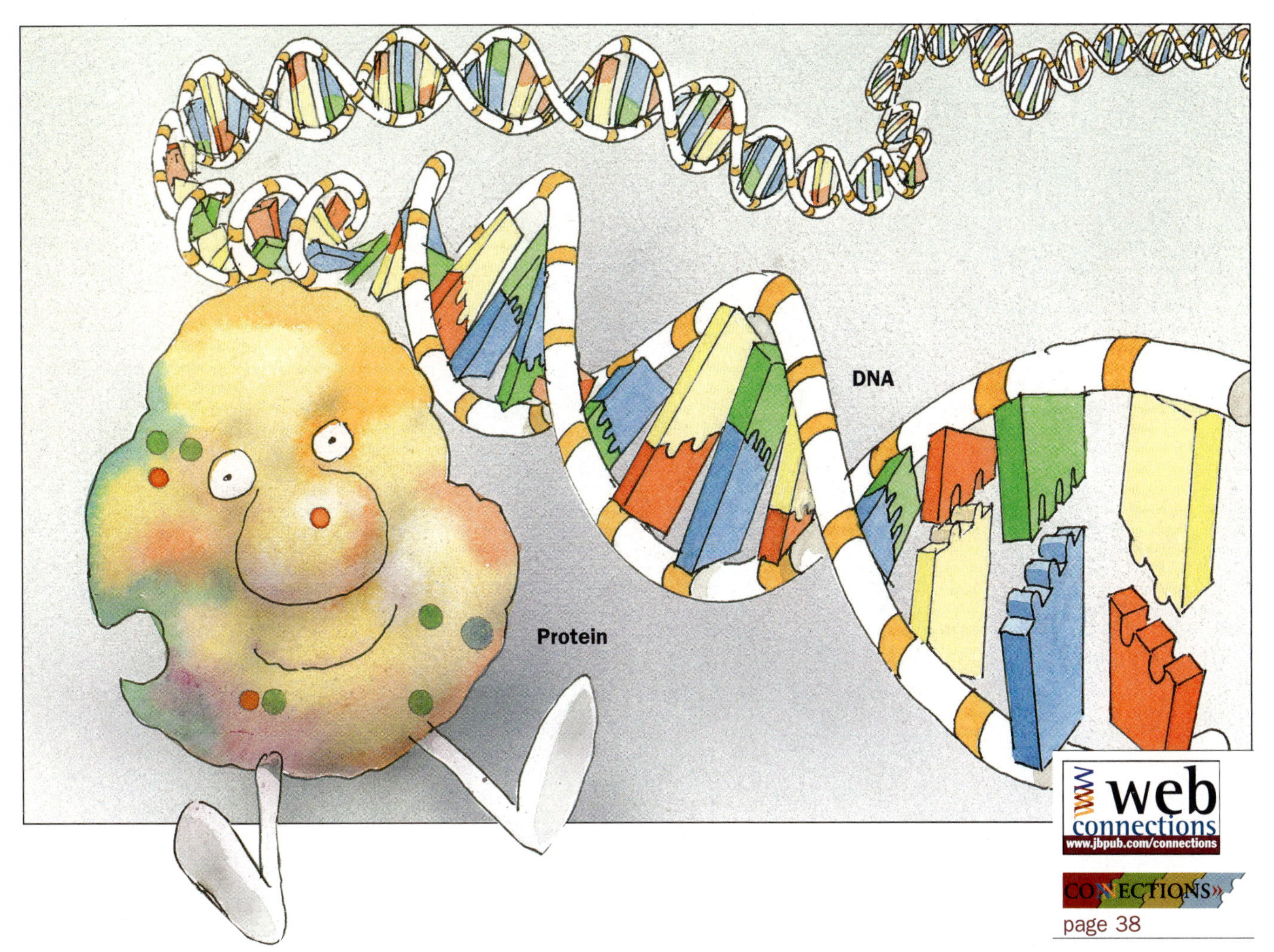

CONNECTIONS»
page 38

The Way Life Works — The Basic Idea

In exploring life's unity, we set out to connect the world of molecules with the world you can see around you.

Our central characters in this story are two chain molecules: One carries the information, the other does the work. To put things simply, you might say that life is played out in the interaction between these two players — DNA and protein, whose relationship can be seen as that between information and machinery.

Picturing the Invisible

Objects the size of atoms and simple molecules, and even DNA and proteins, are truly invisible because, even with the aid of the highest-magnifying light microscope, our eyes can't see them. Although scientists do have other powerful ways of finding out what very small things "look" like, nobody really sees details of molecular structures exactly. Thus we have taken liberties in picturing our principal molecular characters in ways that convey clearly what they do.

We depict DNA as a kind of extended Tinker-Toy structure that readily assembles and pulls apart. Proteins — the working molecules of life — are pictured as somewhat human-like little characters. This distinguishes them — things that act — from other molecules, things that are acted upon. We don't imply that proteins are like people in any other ways — except, perhaps, for a certain obsessive tendency to do the same things over and over again. The protein's affable but blank expression should convey the idea.

PATTERNS

Sixteen Things You Should Know About Life

To see life as a whole — to observe what all life has in common — requires a shift in the way we normally look at things. We must look beyond the individual insect or tree or flower and seek a more panoramic perspective. We need to think as much about process as we do about structure. From this expanded viewpoint, we can see life in terms of patterns and rules. Using these rules, life builds, organizes, recycles, and re-creates itself.

Here we describe sixteen of life's patterns. Most apply to the smallest organisms and their molecular parts as well as to the most complex of us. We make no claim that our list is definitive. We simply invite the reader to think about life from the standpoint of not just what makes each living thing unique and different, but also what it is that unites us all.

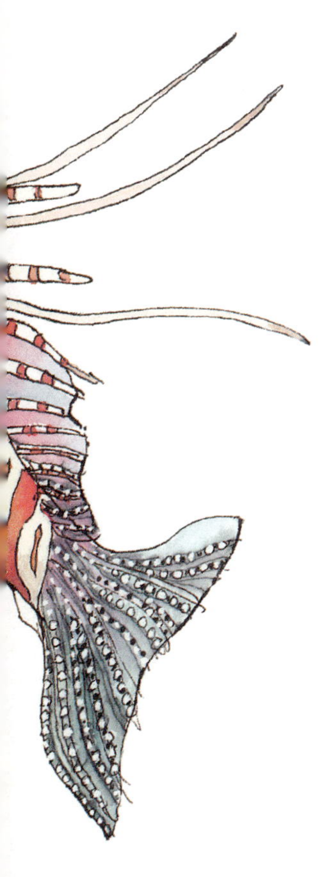

The Sixteen Patterns:

1. *Life Builds from the Bottom Up*
2. *Life Assembles Itself into Chains*
3. *Life Needs an Inside and an Outside*
4. *Life Uses a Few Themes to Generate Many Variations*
5. *Life Organizes with Information*
6. *Life Encourages Variety by Reshuffling Information*
7. *Life Creates with Mistakes*
8. *Life Occurs in Water*
9. *Life Runs on Sugar*
10. *Life Works in Cycles*
11. *Life Recycles Everything It Uses*
12. *Life Maintains Itself by Turnover*
13. *Life Tends to Optimize Rather Than Maximize*
14. *Life Is Opportunistic*
15. *Life Competes Within a Cooperative Framework*
16. *Life Is Interconnected and Interdependent*

1. Life Builds from the Bottom Up

page 39

The Influence of Small Things

"Each living creature must be looked at as a microcosm — a little universe formed of a host of self-propagating organisms, inconceivably minute and as numerous as the stars in the heaven."
— *Charles Darwin*

Early debate about evolution centered around the then-horrifying notion that humans and apes had a common ancestor. But Darwin's idea had far more radical implications: Every individual is a colony of smaller individuals (cells), which are in turn made up of smaller nonliving bits. Further, these smaller bits were the first to develop in our evolutionary history. Occasionally these were usefully incorporated into cells, which, over great gulfs of time, assembled into multicellular organisms. Our ancestors were microscopic, wriggling, squirming creatures similar to what we now call bacteria, whose own ancestors were bits of self-replicating molecules.

Before a single plant or animal appeared on the planet, bacteria invented all of life's essential chemical systems. They transformed the earth's atmosphere, developed a way to get energy from the sun, evolved the first bioelectrical systems, originated sex and locomotion, worked out the genetic machinery, and merged and organized into new and higher collectives. These are ancestors to be proud of!

Given the complexity of the tasks above, we can see why the first multicellular organisms did not appear until the most recent one-eighth of life's duration on earth. So we exist as "corporate elaborations" — composite communities of cells built out of the accomplishments of our one-celled forebears.

Cooperating Communities of Cells

Small communities of cells — like the taste buds on our tongues — work together as an army of specialists. They create a unique structure, with nerve connections to our brain, that allows us to taste the world around us. (The picture at right represents an enlargement of the human tongue.)

Small things are made of yet smaller things.

The bumps on the surface of our tongues, called papillae, contain our taste buds. These, in turn, are formed of clusters of about fifty cells.

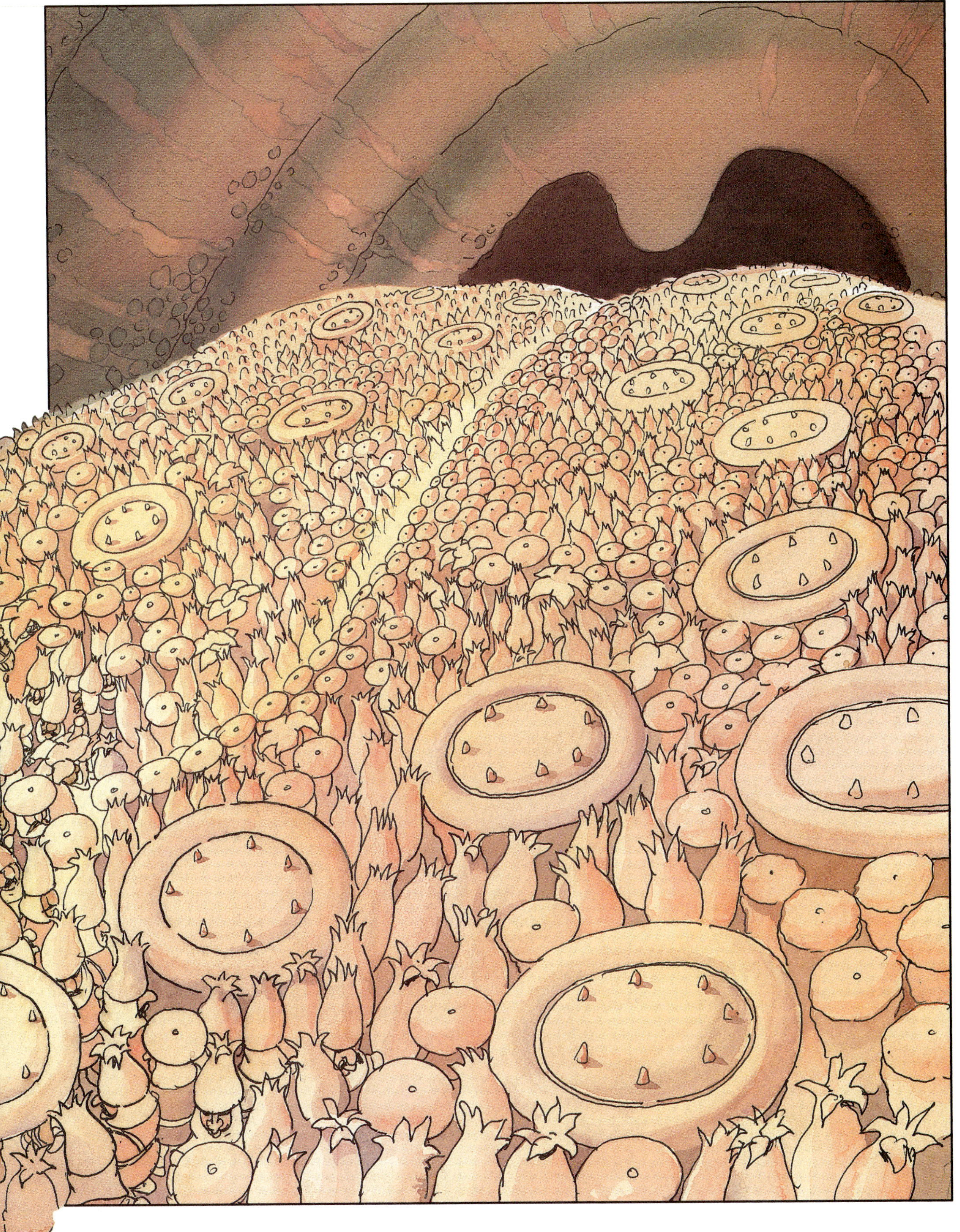

web connections
www.jbpub.com/connections
CONNECTIONS»
page 41

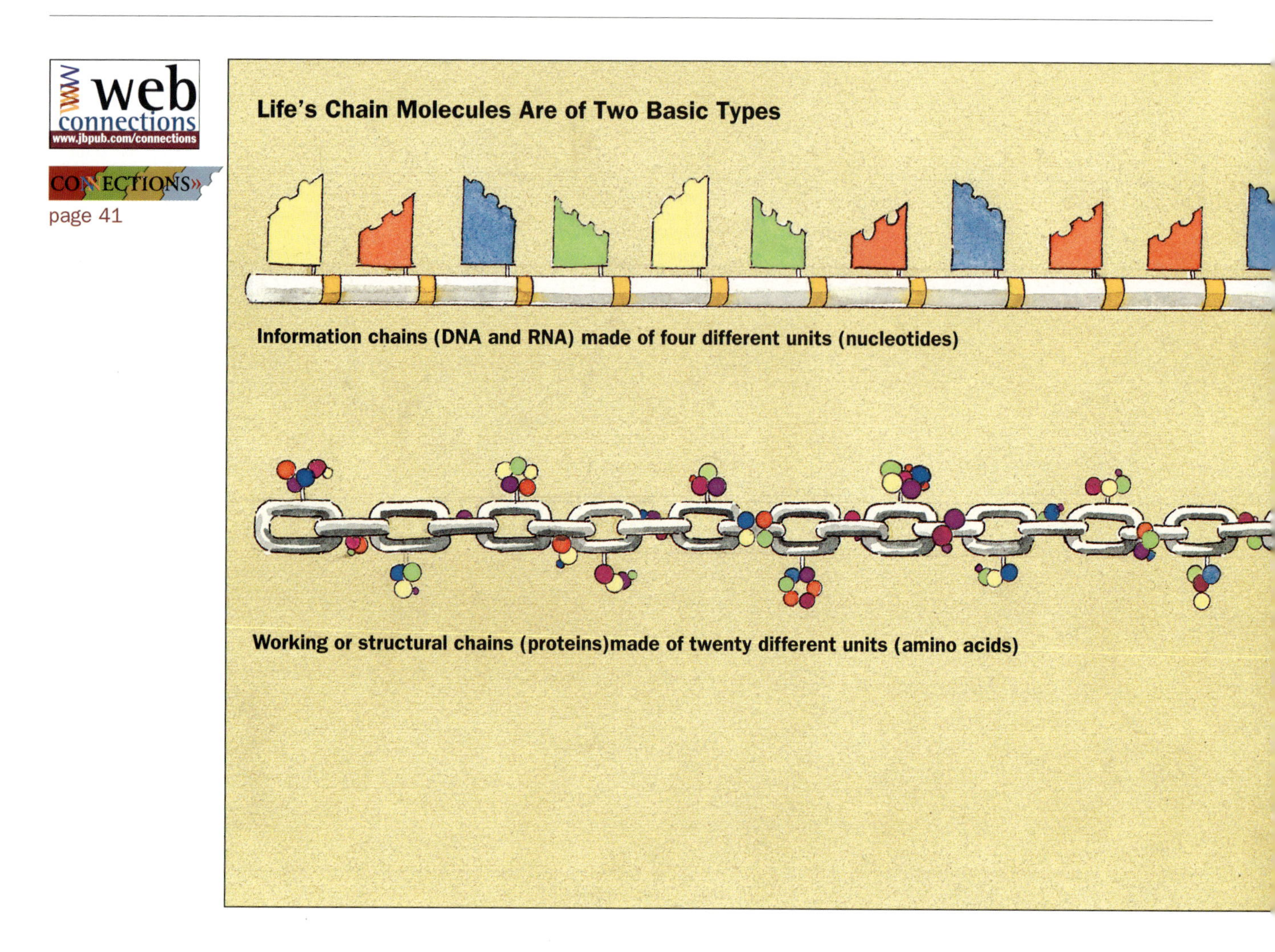

When Difference Becomes Information

*Note: The terms DNA, RNA, nucleotides, proteins, and amino acids are explained in much greater detail in Volumes 3 and 4, **Information** and **Machinery**.*

At the molecular level, life has adopted the chain as its organizing principle. Chains are made of simple units linked together in long, flexible strands. In an ordinary chain, the links are all the same. In contrast, life's chains are molecules containing *different* links. In this respect, the links are the alphabet of life. Letters, in appropriate order, form meaningful words, sentences, paragraphs. Similarly, the sequence of individual links in a chain molecule conveys information.

Chain molecules fall into two main classes: information chains, which store and transmit information, and working chains, which carry out the business of living. Specific lengths of the information chain, called genes, carry the information that becomes specific working chains, called proteins. The two kinds of chains work together in a cooperative loop: Information chains provide the genetic prescription or recipe that is translated into working chains; these in turn make it possible to copy the information chains so they may be passed on to the next generation.

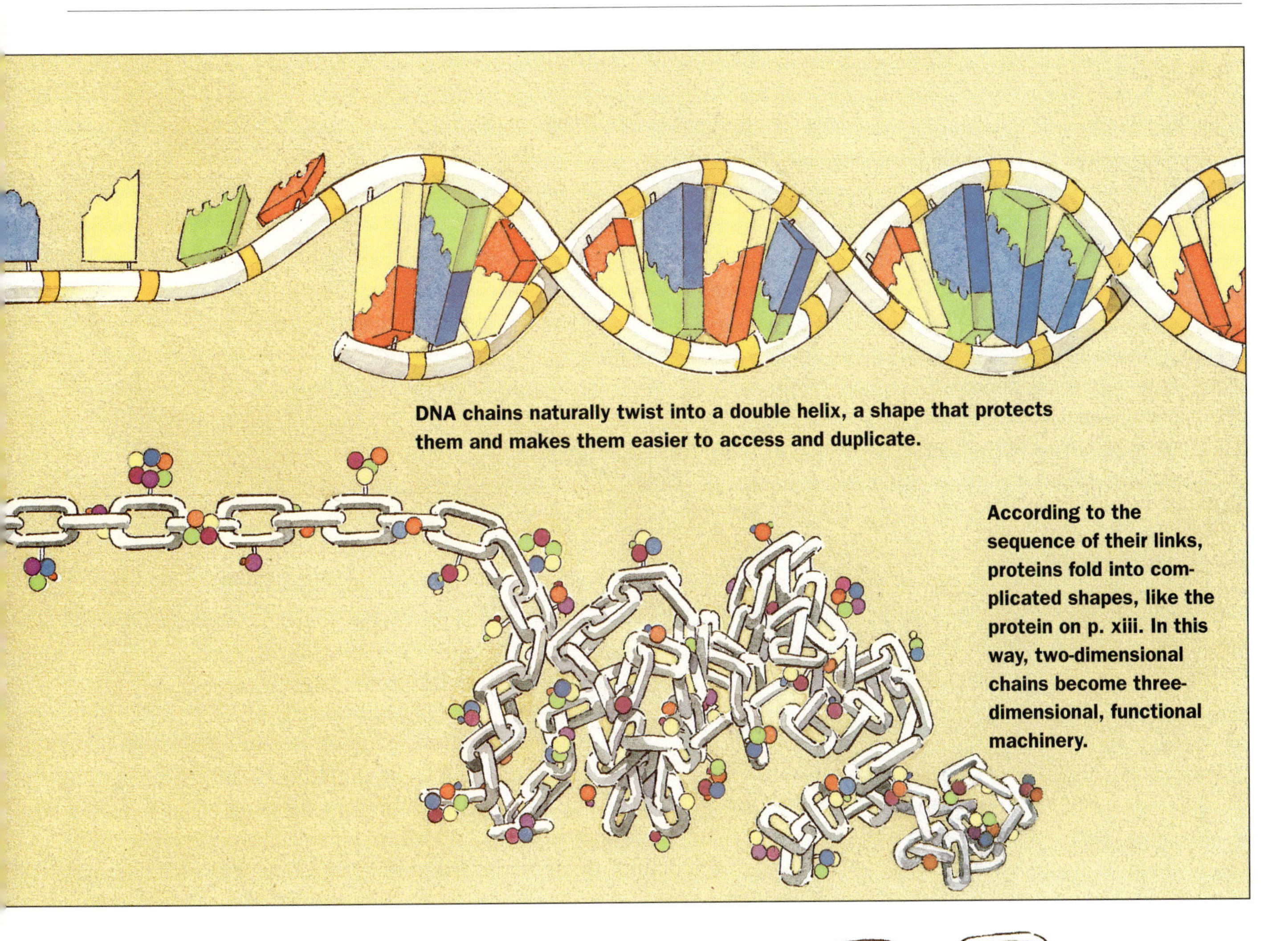

DNA chains naturally twist into a double helix, a shape that protects them and makes them easier to access and duplicate.

According to the sequence of their links, proteins fold into complicated shapes, like the protein on p. xiii. In this way, two-dimensional chains become three-dimensional, functional machinery.

A chain of uniform links is simply a chain —

but a chain of different links can carry information:

—• •—• ••• —•• •— ••— ••• •—

Morse code is a chain of two different units (dots and dashes),

1001100001111010001 10001011110001100100101

computer language is also a chain of two units (ones and zeros), and

Now is the winter of our disco

an English sentence is a chain of twenty-six units (letters).

CONNECTIONS»
page 42

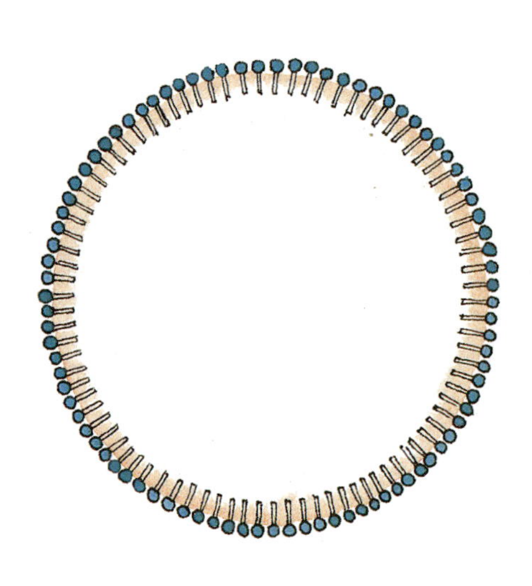

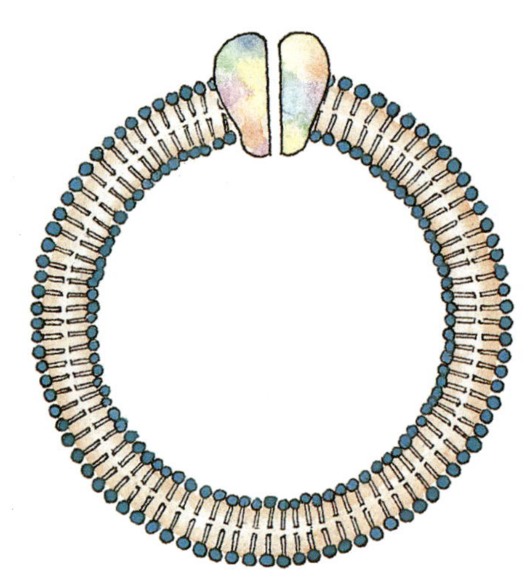

A Cell Membrane

Cellular membranes are formed by combining two layers of regimented phospholipid molecules. On the outer row, the water-liking heads face outward toward the watery surroundings.

On the inner row, the water-liking heads face toward the inside of the cell. The two rows effectively isolate the inner environment. Protein pumps, like the one shown at top, move molecules in and out.

Heads Out — Tails In

When danger threatens, musk oxen gather in a circle — heads and horns to the outside, tails to the inside — sheltering their vulnerable calves in the center. This circle of protection is a memorable analogy for one of life's most fundamental organizing principles — a difference between in and out. Life's chemicals must be kept close together — concentrated — so that they can meet frequently and react readily. To function, the inner environment must maintain a stable level of saltiness, acidity, temperature, etc., different from the outside. These differences are maintained by some form of protective barrier, e.g., a baby's skin, a clam's shell, or a cell's membrane.

The membranes surrounding each of our cells behave something like the threatened musk oxen. The constituent fat molecules have a water-liking head and a fat-liking tail. Heads face outside toward the watery environment beyond the cell; tails face inward. Since the inside of a cell also has a watery environment, a second row of fat molecules aligns itself tail-to-tail with the outer layer, heads facing inward. With this protective structure creating an inside and an outside, plus several pumps embedded in the membrane to move materials in and waste out, life can do its work.

Larger "Membranes"

Bark safeguards the living part of the trunk (usually the outermost ring) from insects, disease, and harsh weather.

The atmosphere helps regulate the earth's temperature as it protects life from the sun's harmful ultraviolet rays.

4. Life Uses a Few Themes to Generate Many Variations

web connections www.jbpub.com/connections

CONNECTIONS» page 45

The Inward Similarity of Outward Diversity

Life hangs on to what works. At the same time, it explores and tinkers. This restless combination leads to a vast array of unique living creatures based on a considerably smaller number of underlying patterns and rules. For example, when cells divide and grow, they do so in a mere handful of ways. New cells can form concentric rings, as they do in tree trunks and animal teeth. They can form spirals, as in snails' shells and rams' horns; radials, as in flowers and starfish; or branches, as in bushes, lungs, and blood vessels. Organisms may display several combinations of these growth patterns, and the scale can vary; but for all life's diversity, few other growth patterns exist.

Life, in striving for the most economical use of space, borrows mathematical rules. For instance, count the branches coming off a stem for a given number of full turns around the stem, and with surprising consistency the numbers of turns and branches relate to each other as in the series 1 1 2 3 5 8 13 21. . . — the so-called Fibonacci series — in which each successive number is the sum of the two preceding it. Thus, in a pine cone, there are thirteen scales for every eight turns. Similar patterns occur in the spirals of florets in sunflowers and daisies, the sections of the chambered nautilus, even the branchings of the bronchial tubes in our lungs. Such similarities in pattern give us some insight into how simple rules, used in different contexts, can produce great variety. From few notes, nature creates many symphonies.

Variations on a Theme

The beetle, with some 300,000 separate species (the world's most numerous order), displays every imaginable color, decorative motif, and proportional distribution of body parts — yet the pattern of relationships that makes the species all beetles is constant.

Different Proportions —
The Same Pattern

Placing these varied fish species within a “stretchable” grid demonstrates that their differences in shape are a matter of proportion. The fundamental pattern is the same.

From d’Arcy Wentworth Thompson, *On Growth and Form*

CONNECTIONS»
page 48

By itself, a sub-robot could never make a complex working part.

For a team of specialists, however, each completing a single step, the task becomes manageable.

Making the Parts That Make the Whole

The business of living requires a lot of information. An organism needs to know how to maintain a constant temperature, how to replace worn-out parts, how to defend against invaders, how to get energy out of food, and so on. It has been estimated that the information a human being needs for all of his or her functions would fill up 15 encyclopedias. It might be many times greater than that but for a strategy life has developed in storing only a certain kind of information. The nature of this information might be best understood by the following analogy: Suppose you decided to build a complex robot requiring millions of individually hand-crafted working parts. Presumably, this task would require instructions for the making of each part, plus instructions for the overall assembly, as well as operating instructions. But now imagine that you had another option: You could acquire the instructions to make several thousand tiny sub-robots, each of which knew how to fabricate one stage of one of the parts. And by working collectively, these sub-robots could assemble and operate the entire robot. In other words, an extraordinarily complex robot would result from complicated interactions among many sub-robots, each of which performs a relatively simple task.

This is the kind of information that life stores in its DNA — in its genes. Genes contain no information on maintaining temperature, defending against invaders, decorating a home, choosing a mate, etc. They contain only information on how (and when) to make proteins. The rest is up to the proteins — life's subrobots.

web connections
www.jbpub.com/connections

CONNECTIONS»
page 50

The playing card metaphor breaks down when we try to stretch it beyond dramatizing the large number of combinations you can get from a small number of variables. Each card in a pack is, of course, different from all the others, as each of our genes is different. And in two packs, matching pairs are identical: two jacks of diamonds, two fives of hearts, and so on. But matching pairs of genes are not always exactly the same. Your gene for the oxygen-carrying hemoglobin protein in your blood may be slightly different from mine. The difference may be trivial, or it can account for poorly functioning hemoglobin in one of us. All the differences in our genes account for why each of us is unique. About one third to one half of all our genes are different in this way.

Mixing Instructions

Nature creates new combinations by exchanging information. The earliest life forms, simple bacteria-like organisms, found a way to inject bits of information into each other — a primitive form of sex. Over time, living things acquired the ability to exchange ever-larger chunks of information, thus evolving sexual reproduction, which is a more elaborate form of information reshuffling.

A very large number of combinations can result from a small number of variables. Shuffling two 52-card decks together produces 4×10^{24} 52-card combinations — a staggering number (4 followed by 24 zeros). But now consider what gene mixing can generate. Think of our genes as equivalent to decks of cards. Each of us has, in all our cells, two decks of cards, one we got from our mother, one from our father. Each gene in one of us is equivalent to a similar gene in another person. But, unlike cards, many matched pairs of genes are not exactly alike. While some of the differences are "silent" — trivial — thousands of them account for why we're different from each other. Also, unlike cards, of which there are only 52 matching pairs, humans have 70,000 matching pairs of genes.

In making our sperm and eggs, we shuffle our mother's and father's decks together, then cut the 104-card deck so each sperm and egg gets half. Then, in mating we combine the two halves. You can see that the number of possible combinations of those 70,000 genes is astronomical. This gives you some idea of the genetic potential for biological diversity.

Mixing Information

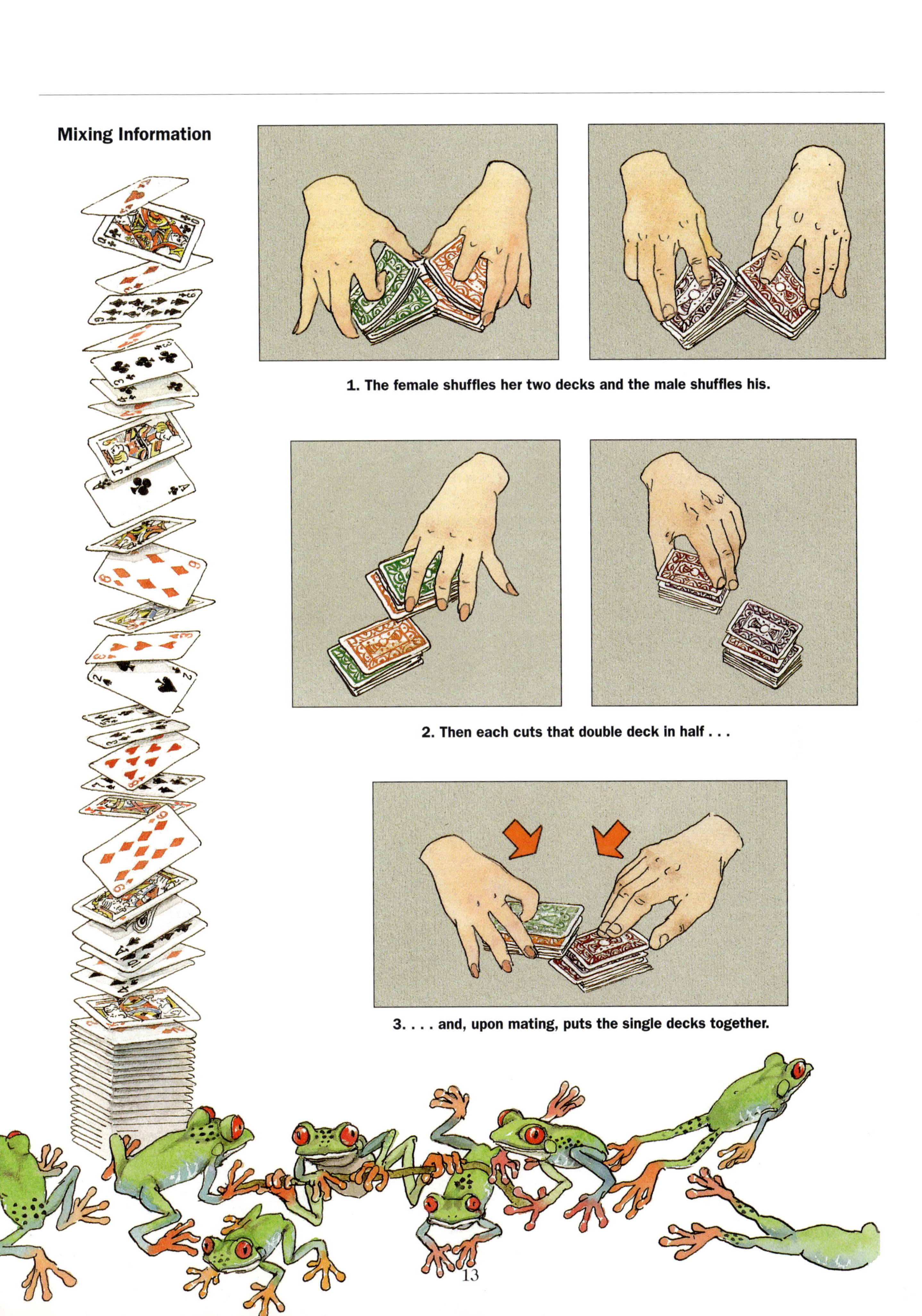

1. The female shuffles her two decks and the male shuffles his.

2. Then each cuts that double deck in half . . .

3. . . . and, upon mating, puts the single decks together.

web
connections
www.jbpub.com/connections
CONNECTIONS
page 51

Size and Surface

Wrinkles and bumps allowed the elephant's ancestors to get bigger. Increasing surface area by creating hills and valleys also allowed organs such as intestines (see p. 58), lungs, and brains to increase their functional capacity while confined within a limited body space.

Accidents Ensure Novelty

When individual cells reproduce, they first make a copy of the information they carry in their genes. Usually this copy is exact, so the information is transmitted perfectly to the next generation. But every so often, cellular mechanisms make errors in gene sequences — sometimes by only a tiny bit. Miscopying even a single nucleotide in a gene, like dialing a single wrong digit in a phone number, alters the gene sequence and therefore changes the piece of information being transmitted. The altered information shows up in the offspring, usually as a defect. But every once in a while, it shows up as an improvement — something that makes the offspring better adapted for survival than its parents.

As an example, take the elephant. Scientists speculate that its early ancestors were small and smooth-skinned. Imagine a copying error in the distant past that jumbled the instructions for the elephant's skin cells, making them assemble into wrinkly and bumpy patterns. It happens that wrinkly skin provides more surface area than smooth skin, a fact of geometry that came in handy for the elephant. Large animals generally have a problem with overheating. A wrinkled skin exposes more surface to the air or water and thereby cools the animal more efficiently. Thus wrinkled skin helped make it possible for the elephant to grow larger and to enjoy the advantages that come with increased size.

As you come to appreciate the evolutionary role of copying errors, it is apparent that calling them "mistakes" oversimplifies. We may, in a larger context, view them as nature's way of introducing randomness, an essential feature of all creative processes.

A Mistake for One Organism Can Be an Advantage for Another

Albinism, a defect in pigmentation, occasionally shows up in many kinds of plants and animals. Most albinos find themselves at a disadvantage in life, since they don't blend into their surroundings, and albino offspring in many species do not survive infancy. Snowy white polar bears, ptarmigans, arctic foxes, and snowshoe hares, however, owe their camouflaging white coloring (and their very existence) to their albino ancestors.

CONNECTIONS»
page 53

The All-Purpose Molecule

Of all the molecules of life, none is so omnipresent as water. Our cells are 70 percent water. Life began in water. When our ancestors arose from the sea to become land dwellers, we brought water along with us, within our cells and bathing them. Most of the essential molecules of life dissolve and transport easily in water.

The most abundant fluid on earth is, happily, the one most suited for encouraging living chemistry.

Water participates in all kinds of chemical reactions. Bounded by water-insoluble membranes, cells owe their shape and rigidity to water. And water provides an inexhaustible supply of the hydrogen ions needed for converting the sun's energy into chemical energy.

What is it about water that makes it so special? The key is its polarity. Composed of a single oxygen atom sharing electrons with two hydrogen atoms — like a head wearing a pair of Mickey Mouse ears — a water molecule looks quite ordinary. While the molecule's overall electric charge is neutral, the oxygen tends to pull negatively charged electrons toward it, leaving the hydrogen "ears" slightly positively charged relative to the more negative oxygen "head."

This means that an ear of one water molecule will form weak bonds with the head of another and vice versa, so that water molecules continuously stick and unstick to each other, thus forming dynamic, evanescent lattices. This self-embracing quality of water accounts for its tendency to remain liquid when most other substances with molecules its size are gases.

Most of life's important molecules are readily soluble in water: they tend to form weak bonds with water as easily as water bonds with itself. The random motion of all molecules, and their tendency to spread out evenly in a solution insures that, once dissolved, they rapidly diffuse throughout the body's watery environment.

Luckily for us, water also has the unusual property of expanding when it freezes, so that the less dense ice floats. This provides an insulating layer that prevents further freezing of our lakes, rivers, and oceans. If water were like most natural materials, whose solid state is denser than their liquid state, ice would sink, and bodies of water in colder climates would freeze solid, making life untenable.

Water's specialness is due to its molecular structure. The two hydrogens (Mickey's ears) have a positive charge, the oxygen, a negative charge...

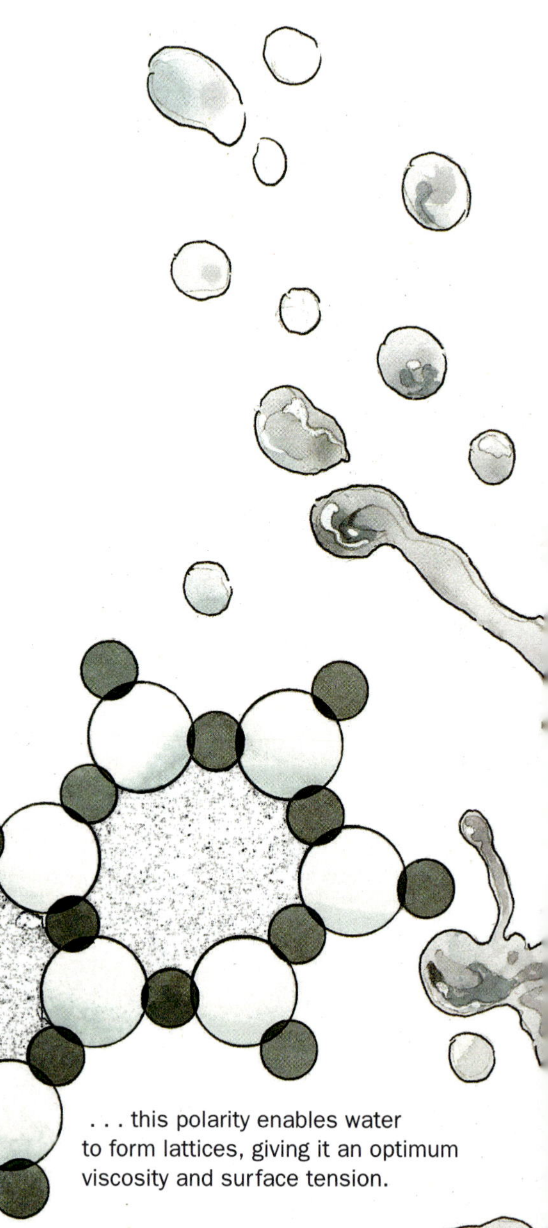

. . . this polarity enables water to form lattices, giving it an optimum viscosity and surface tension.

web
connections
www.jbpub.com/connections
CONNECTIONS»
page 54
SUNLIGHT
CARBON DIOXIDE
WATER
OXYGEN

Plants produce and store sugar for their own consumption. Animals eat the plants or prey on those that do. Bacteria consume the bodies of all. Thus sugar percolates throughout life.

Glucose, life's key sugar molecule, is broken down — metabolized — by living cells, and its parts used to make life's essential molecules.

A Molecule to Burn

Sugars are simple, energy-packed chains of three to seven carbon atoms festooned with hydrogens and oxygens. Life's central sugar is the six-carbon glucose. It is the fuel that drives the engine of life and the basic material from which much of life is constructed. Each year, plants, marine algae, and certain kinds of bacteria convert 100 billion tons of atmospheric carbon dioxide (CO_2) and hydrogens extracted from water (H_2O) into sugar — using energy from sunlight in a process called photosynthesis. The waste product of this massive conversion is oxygen.

Plants and algae and bacteria and animals all "burn" sugar. That is, inside their cells they transform the energy in sugar's chemical bonds into an especially potent form of chemical energy — adenosine triphosphate, or ATP. In this living combustion process, called respiration, sugar's carbons and oxygens are discarded as CO_2 and its hydrogens are linked to oxygen from the air and discarded as H_2O. Thus the very substance of life materializes from air and finds its way back to air. The constantly generated ATP powers all life's work, such as moving, breathing and laughing. Sugar also serves as the starting material for the assembly of the simple molecules — amino acids, nucleotides — from which large molecules are assembled.

Several hundred million years ago, enormous quantities of the remains of trees, plants, animals, and bacteria were buried deep in the earth, subjected to intense heat and pressure, and transformed into coal, petroleum, and natural gas. Much of this material was initially chains of sugar molecules — cellulose and other related chain molecules. So sugar reemerges as the basic ingredient of the fuels that drive the engines of civilization.

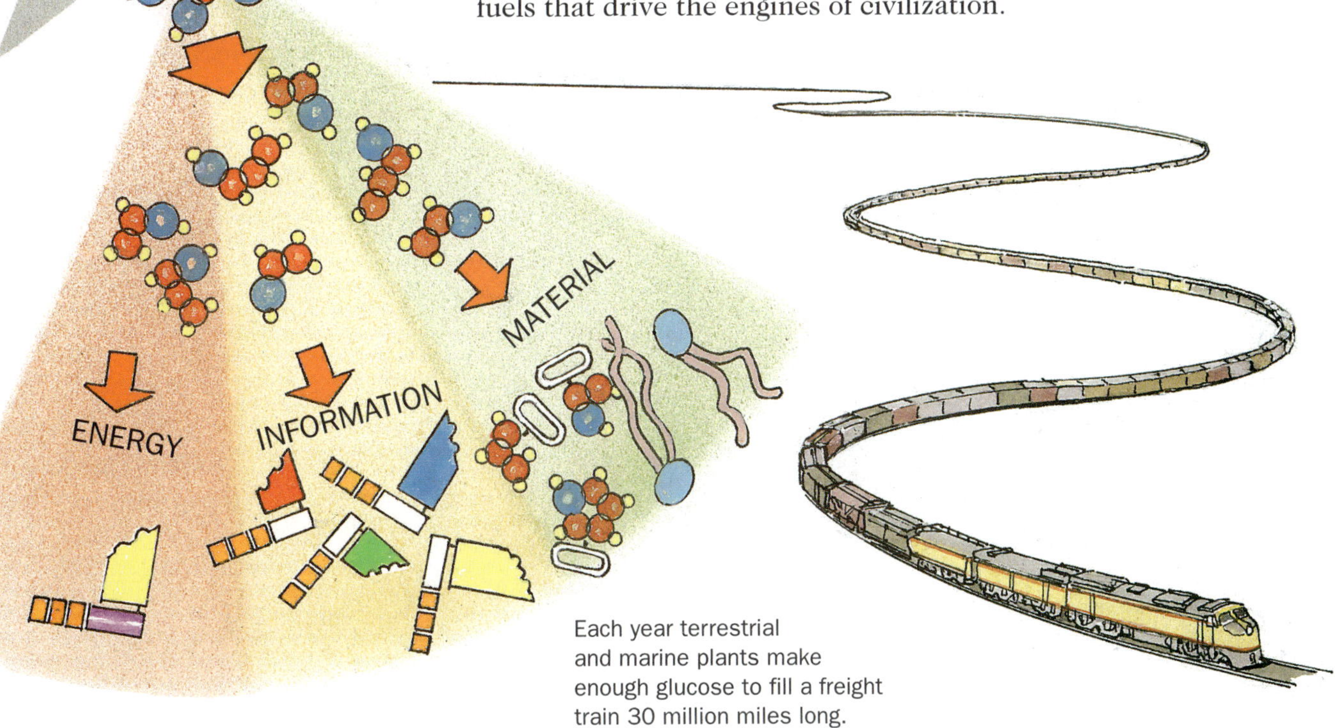

Each year terrestrial and marine plants make enough glucose to fill a freight train 30 million miles long.

web connections
www.jbpub.com/connections
CONNECTIONS»
page 56

The engine's main wheel is turned by steam. A belt from the wheel causes the governor — a spinning ball system — to rotate. The faster the wheel turns, the faster the governor's shaft turns, the farther outward fly the balls. This lifts the disk, raising the lever, and closes the steam input line, slowing the engine.

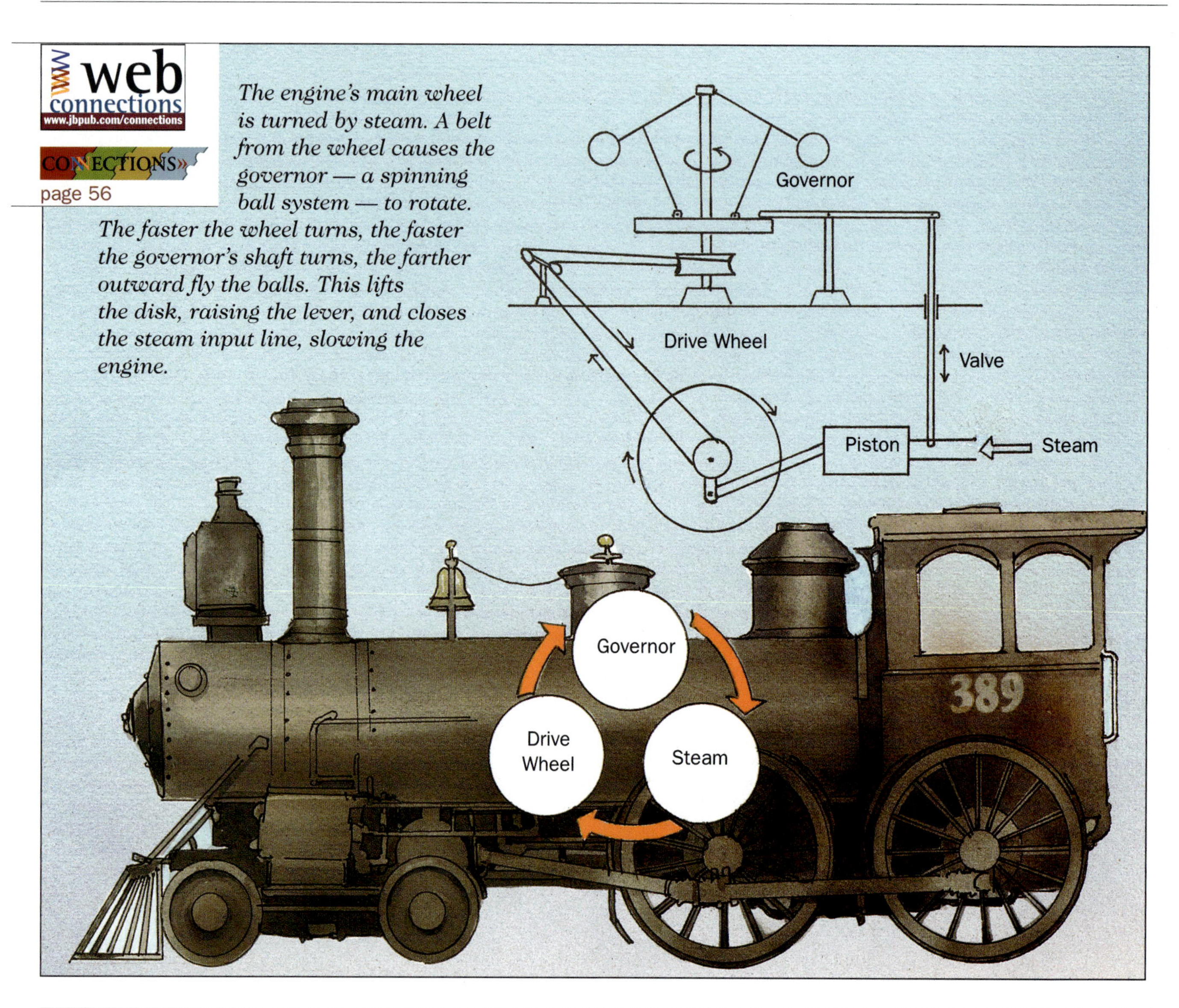

Circular Control

In this simplified steam engine, a fire heats water, making **steam,** which activates a piston, which turns the engine's **drive wheel,** which spins the **governor,** which controls the steam supply. Such a three-component loop passes information from part to part so that the engine is able to self-correct by way of the governor.

A similar self-correcting system comes into play when a protein makes a chemical product. Each protein performs a simple task (e.g., adds a part) in assembly-line fashion. The circular arrangement allows the initial protein to keep track of the overall output. As products either pile up or become scarce, it adjusts the speed of the overall operation. Volume 5, *Feedback,* discusses how it does this.

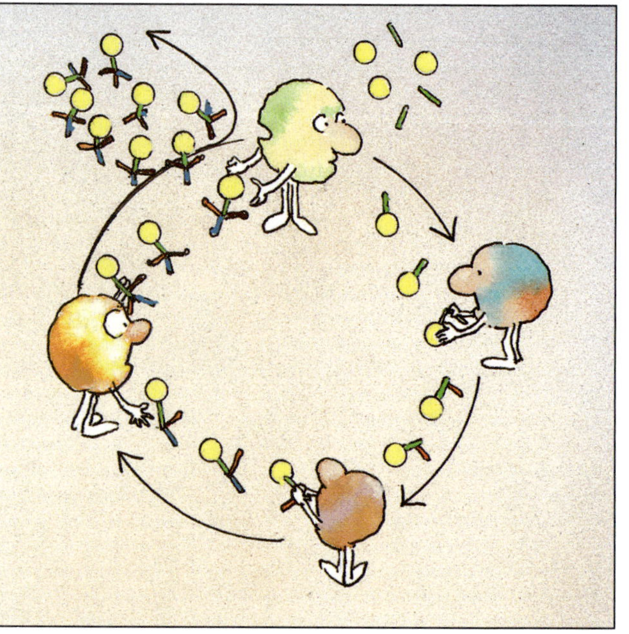

A Circular Flow of Information

Life loves loops. Most biological processes, even those with very complicated pathways, wind up back where they started. The circulation of blood, the beat of the heart, the nervous system's sensing and responding, menstruation, migration, mating, energy production and consumption, the cycle of birth and death — all loop back for a new start.

Loops tame uncontrolled events. One-way processes, given sufficient energy and materials, tend to "run away," to go faster and faster unless they are inhibited or restrained. The steam engine with a governor illustrates the principle: As steam pressure rises, the engine goes faster. The governor, consisting of two rotating arms that lift higher as its shaft spins faster, progressively reduces the steam input; the engine slows; the governor slows; the steam input increases; the engine speeds up. Thus information courses around the circuit to produce action in the opposite direction. The system self-corrects; the parts self-adjust. If such self-generated restraints and inducements occur in small steps, the overall system appears to maintain itself in a steady state.

Every biological circuit, whether a sequence of proteins in the act of consuming a sugar molecule or a complex ecosystem exchanging material and energy, exhibits self-correcting tendencies like those of the steam engine.

Information flows around the circuit and feeds back to the starting point, making necessary adjustments along the way. It's easier to understand how molecular systems assemble into complicated, apparently purposeful organisms when we look at events in terms of multilayered loops of control and creation — and substitute the term "self-correcting" for "purposeful."

Self-Correcting Maneuvers

As an owl tracks a fleeing mouse, she quickly translates the mouse's zigzags into movements of her wings and tail. The owl gets her dinner by maintaining a feedback loop between her eyes, brain, wing and tail muscles and the mouse's movements.

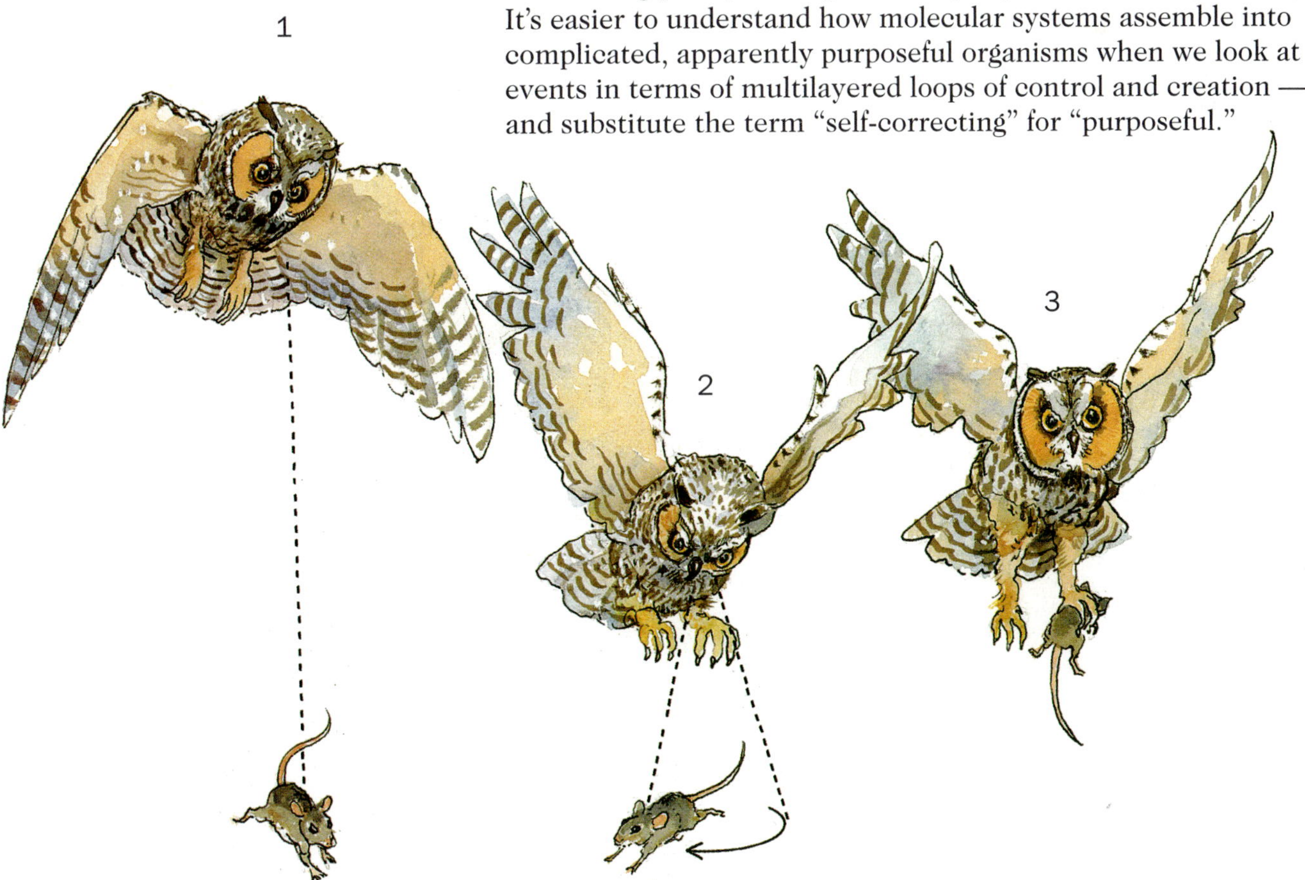

11. Life Recycles Everything It Uses

page 57

A Circular Flow of Materials

For every molecule that the living world makes or uses...

...there exists an enzyme somewhere to break it down.

We humans are unique among animals; we leave behind us a trail of accumulating, unusable products. Everywhere else in the living world, intake and output are balanced, and one organism's waste is another's food or building materials. Waste from a cow circulates from bacteria to soil, to earthworms, to grass, and back to the cow. Crabs need calcium, which they normally get from the ocean, to build their shells. Land crabs, lacking an ocean source, extract calcium from their own shells before discarding them during molting. Hermit crabs save energy by moving into shells cast off by other species, trading up when the shell gets too small.

At the molecular level, key atoms pass from molecule to molecule in a succession of small steps. The end product of one process becomes the starting point of another, the whole train of events bending around into a circle. One creature's "exhale" becomes another's "inhale." Oxygen, dumped by plants as a waste product of photosynthesis, becomes an essential key to combustion in animals' respiration. And the carbon dioxide waste that animals excrete is taken up by plants for sugar-making. From the standpoint of the whole ecosystem, these interchanges occur so smoothly that the distinction between production and consumption, and between waste and nutrient, disappears.

Each generation of living things depends on the chemicals released by the generations that have preceded it.

In a continuous cycle, plants and animals exchange the chemicals necessary for energy and building materials.

web connections
www.jbpub.com/connections

CONNECTIONS»
page 58

Put It Together — Take It Apart

Keeping a living system in a state of high organization necessitates the continuous building and destroying of its parts.

Consider the following dilemma. To exist, life requires organization. Organization requires energy. Life's complex molecules have lots of energy in the bonds that hold them together, but these bonds don't hold together indefinitely. They tend to fall apart — dissipate. Now, a system that is unstable when it's organized has a problem. How can it avoid inevitable breakdown? Living systems have answered this question with an ingenious strategy. Day in and day out, round the clock, organisms routinely take apart their own perfectly good working molecules and then reassemble them. Each day about 7 percent of your own molecules are "turned over." That means virtually 100 percent have "turned over" in about two weeks. In this way, no molecule lingers in your system long enough to "unintentionally" dissipate.

Turnover also provides flexibility. A change in the environment often calls for a switch in proteins. New proteins can be made from disassembled old ones.

In turnover we can sense life's continuous "flow- through" of energy. A high-information/high-energy state must be dynamically maintained by the ceaseless building and destroying, ordering and disordering, of life's parts.

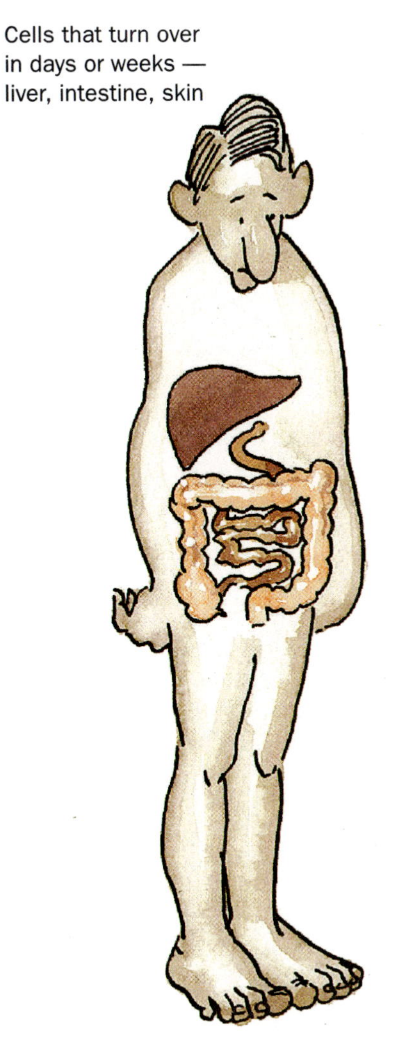

Cells that turn over in days or weeks — liver, intestine, skin

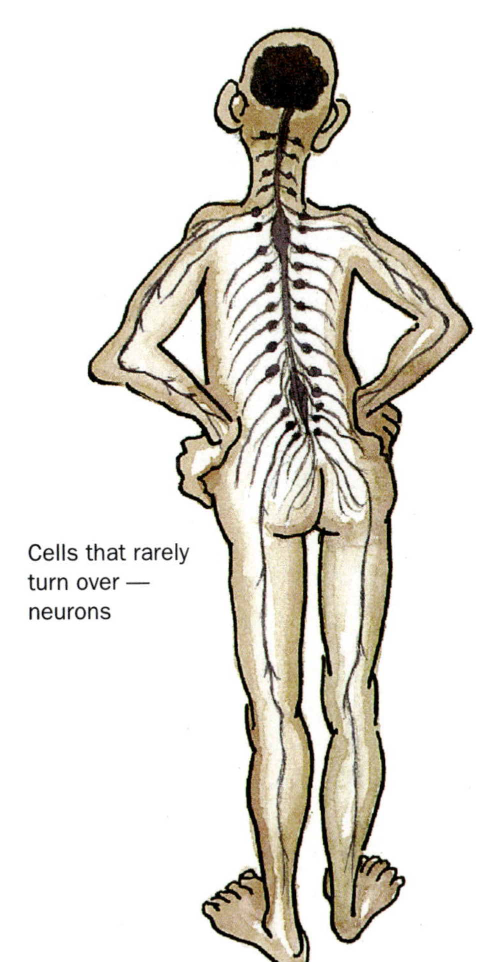

Cells that rarely turn over — neurons

Cell Turnover

Whole cells also turn over; i.e., they have a short or long life, die, and get replaced by new ones.

13. Life Tends to Optimize Rather Than Maximize

CONNECTIONS»
page 59

When Less Is Better

To optimize means to achieve just the right amount — a value in the middle range between too much and too little. Too much or too little sugar in the blood will kill. Everyone needs calcium and iron, but too much is toxic. The rule of optimization generally holds true for minerals, vitamins, and other nutrients the body requires, as well as for behaviors such as exercise and sleep.

At the molecular level, life operates elaborate signaling and management systems to maintain optimum levels. Certain proteins have the ability to regulate precisely concentrations of essential chemicals, shutting down production when optimum quantities have been reached, starting up again when concentrations fall below critical levels.

At the level of the organism, optimizing is an intricate dance involving many interacting parts and values. Deer antlers require an optimum mix of strength, shock absorption, weight, and growing ability (since they must be regrown every year). A change in any one of these variables might adversely affect the others. Something that might make the antlers stronger, like a higher mineral content, might also make them heavier or unable to grow quickly enough. Thus, maximizing any single value (i.e., pushing it to the extreme) tends to reduce flexibility in the overall system, so that it may not be able to adapt to adverse environmental change.

Maximizing can be seen as a form of addiction, in that more leads to more. Occasionally, over generations, an organism may drift from optimizing to maximizing, from adaptation to addiction. The peacock's tail has been cited as an example of the maximizing of one variable trait. If female peacocks choose males who display the most luxuriant tail feathers, the next generation of peacocks will have a greater representation of "big tail" genes. If this process continues unabated, each generation will have a larger average tail size until the tails reach the upper limit of physical praticality. A tail can only grow so large in relation to body size before it impedes a bird's ability to get around. Likewise, a redwood tree can only grow so tall without toppling over; a walrus's tusks can grow only so long without overstraining the animal's neck muscles.

Every once in a while, a sudden change in the environment can catch a species that has drifted too far into maximization and push it into extinction. More often, as the costs of maximization rise, the species self-corrects. Larger-tailed peacocks may be unable to run as fast or hide as well. Because these peacocks are more vulnerable to predators, the survival advantage shifts back toward their smaller-tailed rivals. Thus, life persistently tends toward optimal balance, illustrating one of nature's cardinal rules: "Too much of a good thing is not neccessarily a good thing."

There is, however, one value that life can be said to maximize. Every organism has as its most elemental goal the transfer of its genetic information to the next generation. In this sense, all optimizing of function aims at this ultimate maximization — the survival of DNA.

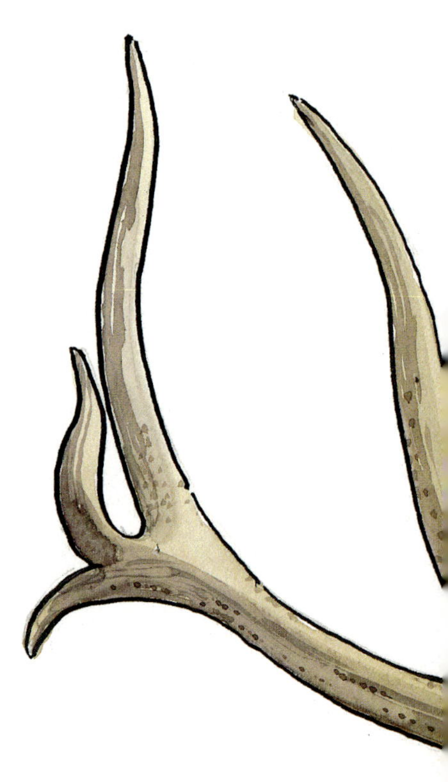

Maximizing to Extinction?

The odd positioning (facing forward) and sheer massiveness (up to twelve feet across) of the Irish elk's antlers suggest they were used for display to attract females, rather than for combat. But in the face of major environmental change — such as the heavy growth of forests — "oversized" antlers might well have contributed to the species' disappearance.

CONNECTIONS»
page 60

Making the Most of What Is

A rotting tree on the forest floor may look like life at a dead end. In actuality, it marks the beginning of an explosive new stage — more varied and bustling than when the tree was alive. Early on, mosses and lichens establish themselves on the decaying surface. Carpenter ants, beetles, and mites initiate a succession of invasions by tunneling through the rotting wood. Fungi, roots, and microbes follow these paths. They in turn become food for grazing insects. Spiders feed on the grazers. Roots of seedling trees and shrubs take hold in the emerging humus as moles and shrews burrow through the soft wood to feed on the newly grown mushrooms and truffles.

The "living dead" tree illustrates not only life's tenacity, but also life's universal tendency to "make do" with whatever is available in its surroundings. Because of this tendency, life flourishes even in the world's harshest places. In Africa's Namib Desert, surface temperatures soar to 150°F, and rain may not fall for three or four years at a stretch. Few plants can survive, yet just under the barren sand live a host of insects, spiders, and reptiles — even a mammal or two. The smallest creatures get moisture from wisps of fog and nutrients from tiny bits of plant and animal detritus blowing across the sands. The larger creatures live on the smaller.

In the arctic ice, 100-year-old lichens grow in temperatures of –11°F. Some antarctic fish have a natural antifreeze running through their blood vessels, enabling them to thrive where others would perish. Tubeworms live in darkness 8,000 feet underwater, depending on minerals streaming from hot water vents on the ocean floor. The world's champion adapters, fast breeding generations of bacteria, can adapt to virtually any environment — from near-boiling sulfur springs to the acid guts of termites. And so on.

Together, over time, the genetic code and the protein structure of all living things permit a marvelous flexibility. Hence, life forms are opportunists. Generations of opportunists don't wait around for the right conditions. They adapt to what is, and they make use of whatever they find around them.

Self-Burial

To avoid winter's harsh dry winds, the mescaxl cactus withdraws completely into the ground.

Growing Toward Darkness

In order to find a tree to climb, the monstera vine must first grow toward darkness. Once it reaches a trunk, it switches strategies and grows toward light.

Adapted to Fire

Resin in the seedbearing cone of the lodgepole pine prevents the scales from opening. Fire not only melts the resin and releases the seed, it also leaves a fertile bed of ashes in which the seedling can take root.

An Invitation to Sex

With the right odor, pattern, and degree of hairiness, the bee orchid entices the male bee into an attempt at copulation. The bee leaves, covered with pollen, to be enticed by another bee orchid.

Living Stones

Lithops look like stones, which helps them avoid being eaten by foraging animals.

Hollow Leaves

Moisture condenses on the inside of the pitcher plant's leaves and is then carried directly to the roots, which need to be kept wet because they are exposed to the air.

Like Rotting Meat

With an evil smell, the Rafflesia flower (actually a fungus) encourages pollination by flies.

CONNECTIONS»
page 61

Strategies for "Fitting In"

1. Every creature acts in its own interests.
2. The living world works through cooperation.

These two statements may appear to be contradictory; they are not. Creatures are self-interested but not self-destructive. Selfish behavior, pushed to the extreme, usually has unpleasant costs. A dominant animal engaging in too-frequent combat may sustain injuries. A parasite may kill its host and have nowhere to go. These self-defeating strategies generally get weeded out by evolution, so that in the long run most everyone tends to adopt some form of "getting along."

Up close, the world looks competitive. From a little distance, its cooperative aspects emerge. A million sperm compete in a winner-take-all race to fertilize a single egg. Sperm production is cheap. Nature can afford to make lots of them to ensure that one is successful. We don't weep for the 999,999 losers. They were part of the system to ensure fertilization, and they did their job. Something similar occurs with predator/prey relationships. Usually predators can take only the smallest, weakest, or most unhealthy of their prey species, leaving the fittest members to survive and reproduce. This may be seen as being competitive at the individual level, cooperative at the group level. (Although we don't suggest that creatures generally *think* in terms of the group.)

Noncompetitors

Although these different species of wading birds feed side by side, they might as well be on separate planets. Each eats a different diet with its unique bill. The fact that each species occupies its own special niche may be taken as evidence for nature's tendency to "get along."

Plants and animals evolved from predator/prey truces among bacteria. The ancestors of chloroplasts and mitochondria (the sugar-making and sugar-burning components of plant and animal cells, respectively) originally acted as small predators, invading larger bacteria. They exploited but did not destroy their host. Such "restrained predation" is a recurring theme in evolution, and in it we see the beginnings of cooperation. In time, the host developed a tolerance for the invaders, and each began to share the other's metabolized products. Eventually they became full-fledged symbionts — i.e., essential to each other's survival. This progressive cooperation set the stage for all higher life forms. The lesson, as biologist Lewis Thomas has stated, is not "Nice guys finish last," but rather "Nice guys last longer."

From Predation to Cooperation

A parasitic mitochondrion invades a larger bacterium.

Many generations later, invader and host begin to share metabolized products.

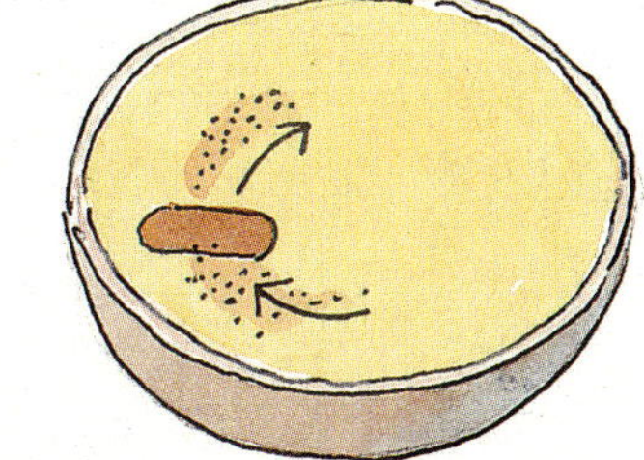

After many more generations, they've come to need each other.

Ritualized Aggression

Animals compete to establish dominance. Such fights rarely result in serious injury and frequently involve only "displays." This can be seen as cooperative behavior.

web connections
www.jbpub.com/connections
CONNECTIONS»
page 62
Nudibranchs are born defenseless but acquire a protective toxin by eating poisonous anemones and incorporating them into their spines.
Parrot fish nibble away at the reef while grazing on algae. In the process they excrete calcium as a fine sand. Each fish produces thirty pounds of sand per year, playing an important role in building beaches.
Pink algae use the reef as a secure place to grow. At the same time, they contribute mightily to holding the reef together by secreting a limey "glue."
Crabs encourage sponges to grow on their backs. A good sponge growth discourages octopuses from eating the crab.
Sea squirts carry tiny creatures called nephromyces in their kidney-like organs. Inside the nephromyces live special bacteria. Both the nephromyces and the bacteria appear to be useful in recycling nitrogen for the sea squirt.

A Network of Interactions

The stony coral, a pea-sized animal that resembles a miniature flower, might easily go unnoticed were it not for the tiny limestone cup it secretes for its home site. As the multiplying coral add on their cups, they form vast apartment complexes — the largest life-made structures on earth. Pink algae, taking hold in the crannies, "mortar in" the loose and broken sections with a limey secretion of their own. Turtle grass, sea fans, sponges, and mollusks attach themselves to the reef surface. Moray eels take up residence in the dark crevices. Starfish arrive to feed on the coral, and triton conches feed on the starfish. Hundreds of species of fish — some grazers, some predators — move in, along with crabs, octopuses, shrimp, and sea urchins. Competitive and cooperative relationships emerge.

Damselfish flit with complete immunity among the poisonous tentacles of the large sea anemones. Crabs place sponges on their backs where they grow and act as a protection from octopuses. Cleaner fish and shrimp remove parasites from predator fish, even entering their gills and mouths with complete safety. Algae live comfortably inside the coral's cells, and large sponges offer housing to thousands of minute creatures.

Look at the coral reef as a multilevel, integrated system. Ultimately, everything in the reef connects with everything else. The survival of the reef shark is closely tied to the survival of the coral polyp, even though the two may have no contact and certainly no awareness of each other. What survives and evolves are patterns of organization — the organism plus its strategies for making a living and for fitting in. Any successful change of strategy by one organism will create a ripple of adjustments in the reef community. Called coevolution, this is the kind of creative force at work everywhere life has taken hold.

Reef-building coral polyps harbor tiny algae within their cells. The algae promote the coral's growth and receive carbon dioxide and nutrients in exchange.

Cleaner fish live safely in the mouths and gills of larger fish, removing parasites.

A Singular Theme

◀Concepts, p viii

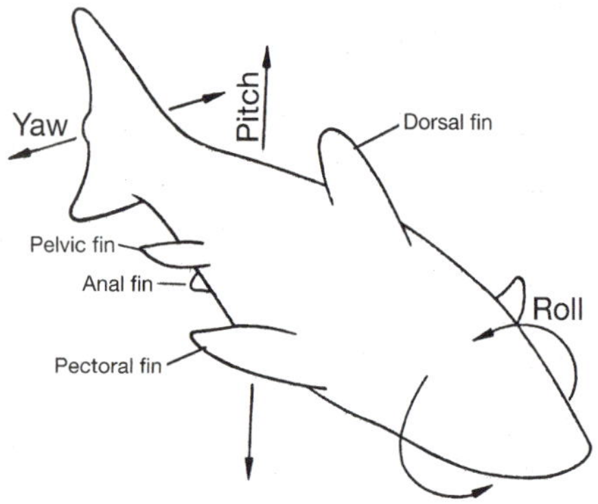

"... the proportions differ, but the pattern is remarkably similar."

Yaw
Pitch
Dorsal fin
Pelvic fin
Anal fin
Roll
Pectoral fin

Navigating the River of Change—A Different Kind of Similarity

Animals that must swim rapidly through the water in order to feed and breathe are more successful if they have a body shape that deals efficiently with the kinds of forces moving water exerts on them. Swiftly moving water (or even prey) may shove them from side to side (yaw), throw them up or down (pitch), or roll them over. It seems that the best shape a marine animal can have to resist these forces is the one you see top left. It is not a coincidence that most airships (which fly through another kind of fluid) have many of the same features.

Over the course of hundreds of thousands of years, groups of animals, very different in origin, like sharks, reptiles and mammals (below left), discovered that the ocean was a relatively safe and nutrient-filled environment. By chance, certain individuals in succeeding generations of these ancestral animals were born with features approaching the useful shape shown above. These individuals swam and hunted and survived and bred better than their siblings. Their descendants did the same, and eventually came to look very similar. The patterns their bodies developed converged toward similarity. Their similar-looking tails and fins had different origins and are analogous, not homologous.

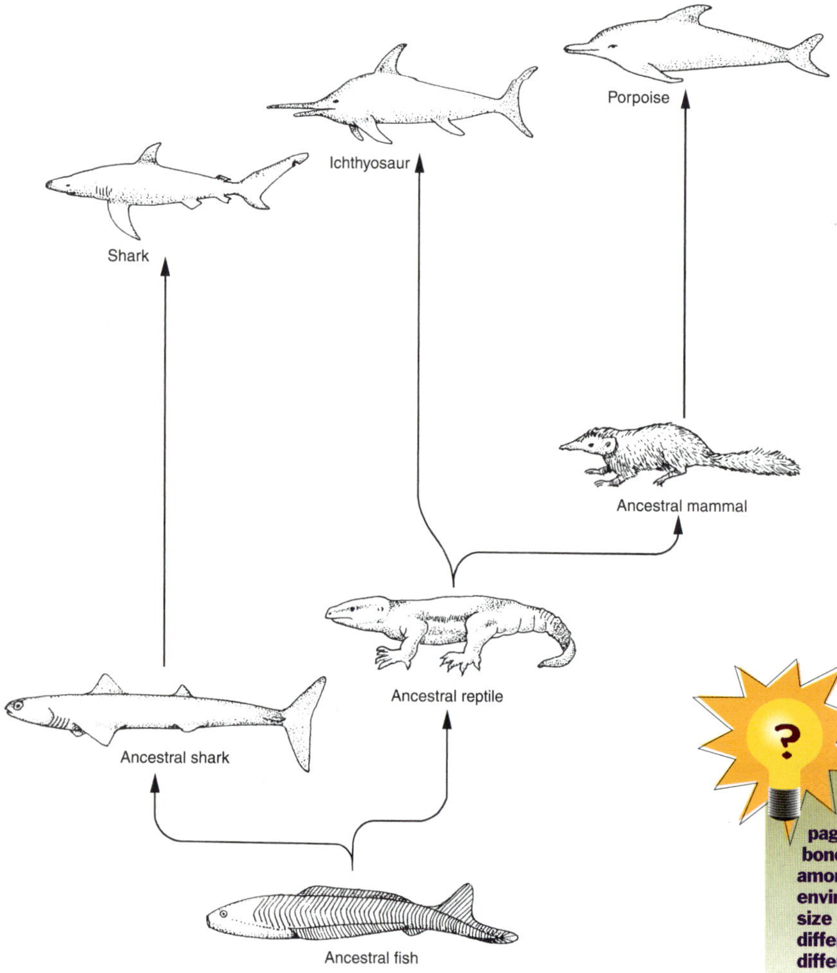

Porpoise
Ichthyosaur
Shark
Ancestral mammal
Ancestral reptile
Ancestral shark
Ancestral fish

Look carefully at the skeletal forelimbs of frog, lizard, bat, bird, human, cat, and whale on the next page. See if you can identify and label the homologous bones in each. Describe some of the useful differences among these structures that make them best suited for the environment and needs of that organism. Can you hypothesize as to why the mass of bones of the bat and the whale differ so much? What is an explanation for the structural differences between the bat's forelimb and the bird's?

From Atoms to Cells – Comparative Sizes

◀ Concepts, p xii

"... the cell's infrastructure, the equivalent of its roads ... libraries."

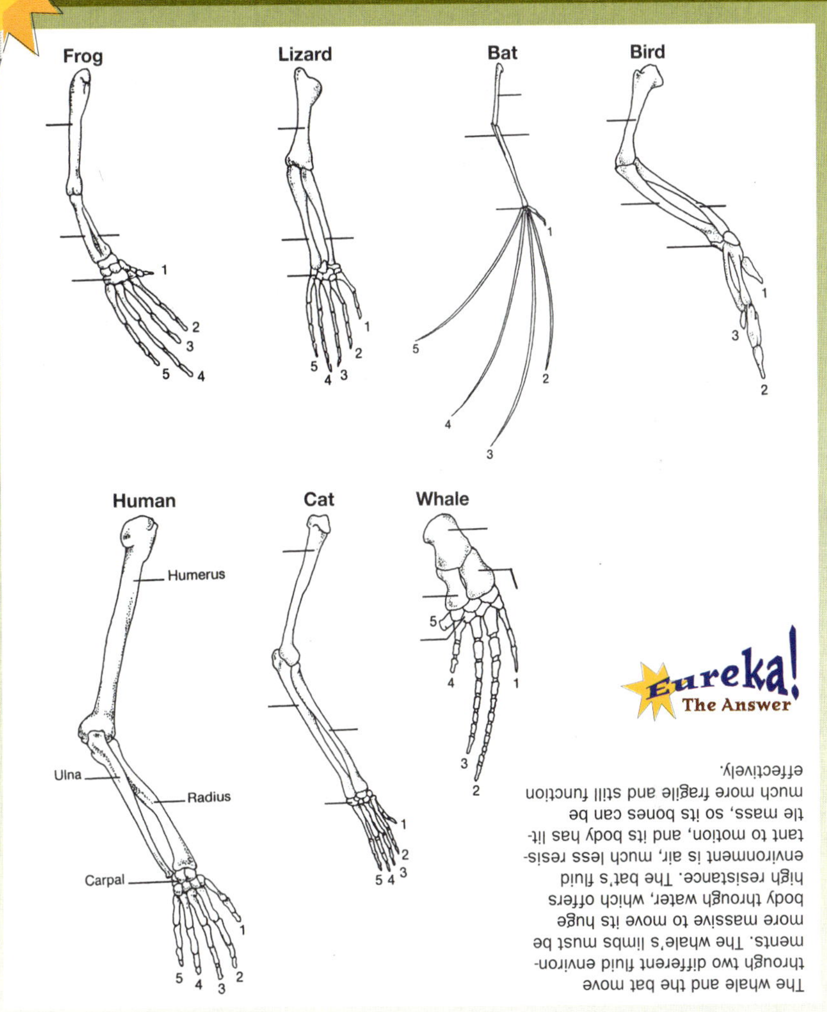

Focus on the Minuscule

All the pictures you see on pages xii–xiii of atoms and molecules, chain molecules and molecular structures, including a bit of a mitochondrion, depict things that until this century, were invisible and thus unknown to us. In fact, it was not until the mid 1600s that the first evidence of the existence of things smaller than the unaided eyes could see was gathered.

In the 1660s, Robert Hooke was the first to report using a magnifying lens to view a thin slice of cork. The densely packed empty chambers he saw, and carefully drew, reminded him of monks' cubicles, and so he called them "cells." This was the first microscopic observation of cells—the basic structural unit of living things.

During the next three centuries, many improvements were made to the basic microscope (called the light microscope), and people were able to observe a previously unexplored universe of microstructures and microorganisms. Cells and many details of their surfaces and interiors, bacteria and their moving parts and millions of other tiny living things, are vividly observable under a light microscope. But the light microscope has its limits. To get a closer view, a new technology had to be created. To communicate the sizes of the minuscule things observed , a new system of measurement had to be devised.

All the units we use for measuring the sizes of these minuscule structures are subdivisions of a meter—the standard unit of length used in the metric system (or International System of Units). A centimeter is 1 hundredth of a meter (10^{-2}), a millimeter is 1 thousandth (10^{-3}), a micrometer is 1 millionth (10^{-6}), and a nanometer (10^{-9}), the unit we use on pp. xii and xiii, is 1 billionth of a meter.

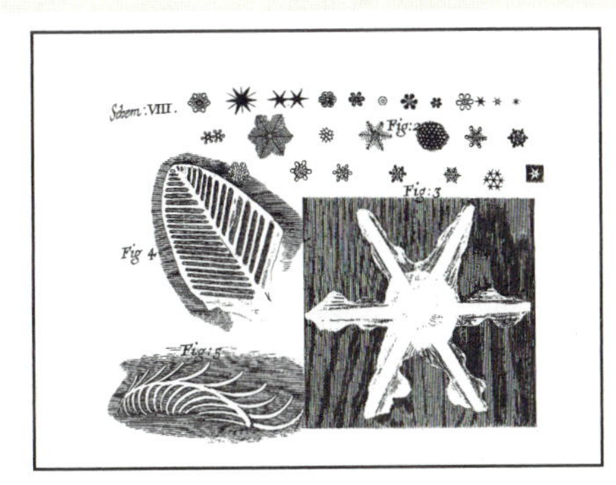

▲ Hooke was so enchanted by the new world of tiny things that he studied and observed anything and everything.Among his "Descriptions of Minute Bodies Made by Magnifying Glasses," were these snowflake structures at four different magnifications.

▶ This is a single strand of DNA magnified as much as is possible with current technology. Magnifying a 12-inch length of string this much would yield an image twelve miles long. We can see that DNA is a long molecule, but we get no information about the details of its structure.

The light microscope, which works by magnifying and focusing the image formed when light passes through an object, cannot distinguish objects smaller or closer together than the shortest wavelength of visible light. (Visible light is just that—the light we can see, and its shortest wavelength is small indeed—about 200 nanometers or nm). A protein molecule, like the one shown at different scales in all the drawings on pages xii and xiii, is only about 4 nm, and so is invisible through a light microscope). The transmission electron microscope or the scanning electron microscope, uses a beam of electrons controlled by electric or magnetic fields. With such a microscope it is possible to see details of cell surfaces and interiors invisible with a light microscope and to discern at least the rough shapes of large molecular structures such as ribosomes.

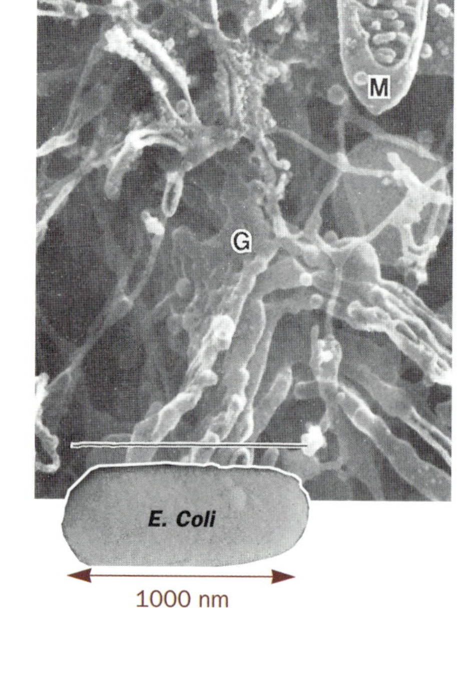

"Magnified 1 million times"

◀ Concepts, p xiii

Comparing Sizes

This scanning electron micrograph of part of the inside of a cell (left) shows a piece of a mitochondrion (marked M). The scale line (a bacterium) is 1000 nanometers, or 1 micrometer. The mitochondrion is about the size of the common bacterial cells *(E. coli)* that live in our intestines. Ribosomes and proteins are far smaller, almost invisibly so (Scale 3, p. xiii,) even to the most powerful microscopes. We can only visualize the details of these complex, interlaced chains of molecules through the mathematical magic of computer models, using data from x-ray diffraction (see p. 38 Picturing the Invisible).

▶ A computer-created model of the oxygen-carrying protein hemoglobin, a chain molecule contained in all of our red blood cells. Each tiny sphere represents an atom of carbon, nitrogen, hydrogen or iron. The atoms are bonded together in amino acid molecules, which in turn are strung together into chain molecules, which fold and twist into very specific molecular structures, of which four then join to form a multiple-chain molecule called a globular protein.

"... the cytoplasm, where most of the cell's active processes occur."

◀ Concepts, p xiv

Exploring the Cell

How many bacteria can live on the point of a pin? Quite a few. The rod-shaped bacteria in the picture at the right, each a single cell, are about the size of the mitochondrion you saw on the preceding page, anywhere from 1,000 to 1,500 nanometers long. A typical animal cell is much larger; its nucleus (the library), a portion of which you see below in a cell broken open by freezing and then fracturing it, is roughly ten times larger in diameter than the bacterium.

Another view of the cell (right) on a similar scale of magnification, but using a different magnification technique, shows part of the nucleus (N), and more of the cytoplasm's contents, such as Golgi bodies (protein-packaging factories, G) and vesicles (V).

The smallest living cells are about 1 micrometer, or 1 millionth of a meter, in size. The largest cells are eggs; an ostrich egg is a single cell! Some animal nerve cells are incredibly long, reaching from the spinal cord to the foot, although they are so thin that you would still need a microscope to see them.

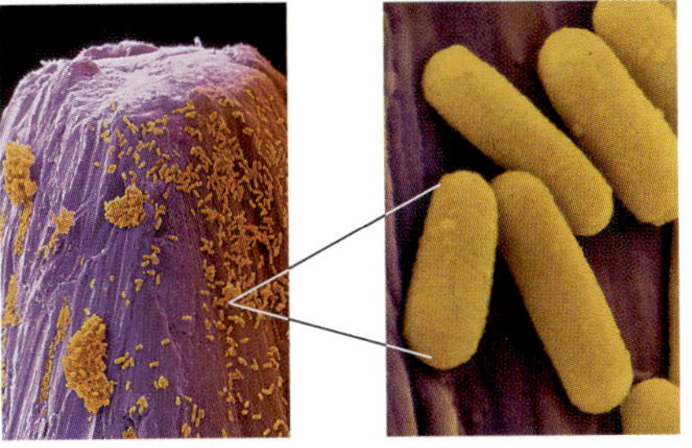

▲ A magnified pin point with a population of bacteria.

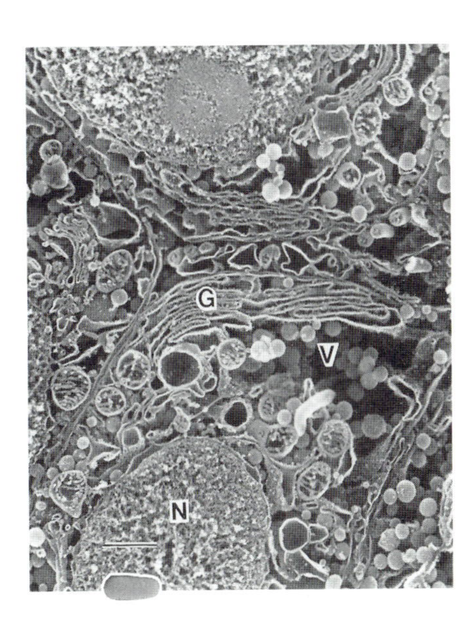

▲ This view allows us to see many of the cell's molecular structures, or organelles.

◀ The rounded shape in the center of the picture to the left is the cell's nucleus; it is covered by outer and inner nuclear membranes (ONM and INM) and surrounded by cytoplasm, in which float various organelles. PM (plasma membrane) marks the double membrane that envelopes all of the cell's contents.

Picturing the Invisible

◀ Concepts, p xvii

"... science does have other powerful ways of finding out ..."

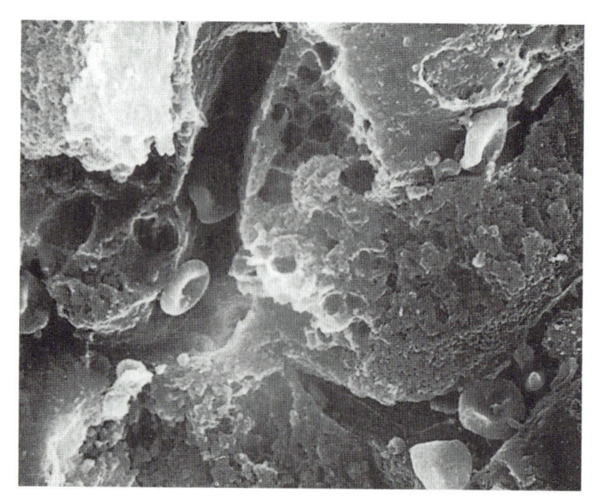

▲ Here inside the liver of an animal, we can see a dramatic three-dimensional landscape of liver cells, connective tissue, and blood cells.

Not only do microscopes allow us to see tiny things, they allow us to see them in their natural "landscape," in relation to the other structures around them. If we want to go deeper and study the structural details of the individual protein molecules that do the cell's work and are critical parts of cellular structures, we must resort to a different approach. First, we must "isolate" the protein molecules—free them of all of the associated materials that surround them. The protein molecules are crystalized, so that they stack regularly in a three-dimensional lattice. Then, we apply x-ray crystallography, which allows us to "see" the molecules at a wavelength of 0.1 nm (the diameter of a hydrogen atom!). This technique contributed to Watson and Crick's discovery of the double helix structure of DNA and Perutz and Kendrew's description of the structure of hemoglobin—all Nobel-Prize-winning achievements.

X-rays, like visible light (and the microwaves in your oven, for that matter), are a form of electromagnetic radiation, but their wavelength is much smaller: 0.1 nm. If a beam of x-rays is focused on a protein crystal, most of the rays pass through, but some are deflected, or scattered, when they hit the atoms.The x-ray will produce a diffraction pattern—a pattern of exposure spots on a photographic film placed behind the protein sample. Regularly repeating atoms in the crystal structure deflect the x-rays at certain angles, creating, on the film, spots whose density and spacing correspond to the density and spacing of the atoms.

Collaboration among many scientists has combined information from x-ray diffraction, electron microscopy, and other technologies. This information, entered into enormous data banks, is the basis for accurate computer models of various, amazingly complicated protein molecules. Now we truly can visualize the invisible.

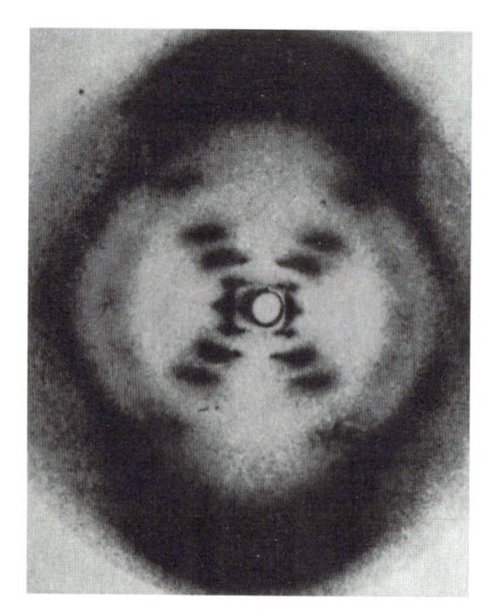

◀ In this diffraction pattern of DNA captured by Rosalind Franklin and deciphered by Watson and Crick, the distance between spots forming the X indicates the distance between turns of the DNA's helix. The X is a reliable indicator of a helical (corkscrew-like) molecular shape.

1. Life Builds from the Bottom Up

◀ Concepts, p 2

"Our ancestors were microscopic, wiggling, squirming . . ."

◀ Volvox, a spherical colony of many single cells, rolls through the water of a pond. Inside the sphere grow daughter colonies, which from time to time break out to spin away on their own.

From Bottom Up to Top Down

The building of life from the bottom up, (i.e., from single-celled creatures into multicelled creatures), suggests a one-way evolutionary direction—from the simple to the more complex. This, however, is only part of the story. As multicellular creatures evolved, they created new environments for the already existing simpler creatures. For example, the unicellular bacteria residing in the guts of all animals live in a mutually evolving dance with their larger hosts. They provide important benefits, including making the host's gut inhospitable to disease-causing organisms and producing necessary substances such as vitamins. In some animals, bacteria also secrete powerful digestive enzymes that break down food to prepare it for the host's own digestive resources. Though the ancestors of these helpful microbes clearly existed before animals, it is highly probable that their hosts contributed to the direction of their later evolution.

This is also the case with parasites, organisms that do not benefit their hosts. They follow a simple rule: "Why make a product yourself when you can easily get it from someone else?" Bacterial and viral parasites, in particular, must have evolved *after* their hosts. And it's likely that many parasites have actually become simpler than their ancestors were.

Finally, consider organisms whose evolution humans have genetically engineered: bacteria that eat oil or attack crop pests. Beyond demonstrating our growing ability to manipulate nature, such creatures exemplify the ongoing worldwide coevolution of micro and macro enviroments.

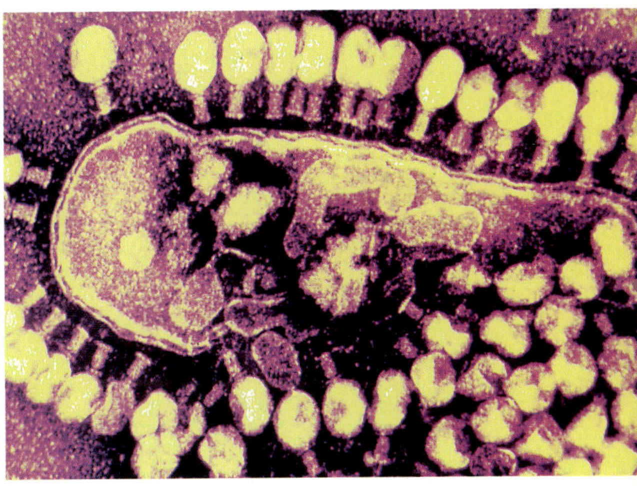

▲ The ranks of balloon-shaped objects lined up along the outer edge of this bacterial cell are viruses. About 30 minutes before this picture was taken, they injected all of their DNA chain-molecule contents into the bacterium. The bacterium's molecular machinery leapt into action and translated the information in those DNA molecules into new viruses, which you can see bursting through the cell membrane at the lower right. Viruses are so simple that they cannot exist without using another organism's molecular machinery to travel and reproduce.

?

The colony of cells called Volvox shown above right can be made up of as many as 50,000 individual cells, each with two whip-like propellors (flagella). The cells are held together in a gelatinous sphere, not actually connected to each other as are the cells of your tongue, let's say, or those of any multicellular organism. Still, the flagella move in a coordinated way to roll the colony through the water, and the colony does seem to have a forward and backward orientation, as well as an inside where new colonies form.

Why do you suppose such colonies might be more likely to survive and reproduce than might free-living single cells over time. What might be the advantage of a colony's remaining just a colony, rather than evolving into a multicellular volvox organism?

Eureka! The Answer

In a fairly calm water environment (where the spheres are likely to remain whole), a colony has an advantage over single cells to have, instead, daughter colonies that develop in a relatively protected place. Remaining a colony might also be an advantage, because a bite from a predator would just downsize the colony, not kill it.

"So we exist as corporate elaborations . . ." ◀ Concepts, p 2

Rebuilding the Corporation

A sponge is perhaps the simplest "corporate elaboration." Its cells function like a cooperating group of individuals. Unlike other multi-celled creatures, sponges have no true tissues (groups of differentiated cells that work in concert, as in taste buds or muscles), let alone organs. Sponges have eight to ten different types of cells, which cooperate to maintain constant water flow through the pores, to trap food, to create fibers and mineral structures that maintain the sponge's shape, and to transport nutrients and wastes.

Probably because of their relative simplicity, sponges regenerate easily: chop a living sponge into pieces and each piece will become a new sponge. Even pressing a live sponge through a fine sieve won't kill it. Deposited on a culture medium, the tiny fragments will begin to migrate and clump together in mounds that eventually become miniature new sponges.

▲ A colony of a very ancient type of tropical sponge. Notice the tubular, porous structure, ideal for enhancing water flow. Specialized cells with long appendages called flagella create currents that pull nutrient-filled water in from outside and up through the tube. Each individual cell exchanges nutrients and waste with the outside world. The human intestine and trachea (both tubular organs) are also lined with cells that perform the same kind of absorbing and transporting function.

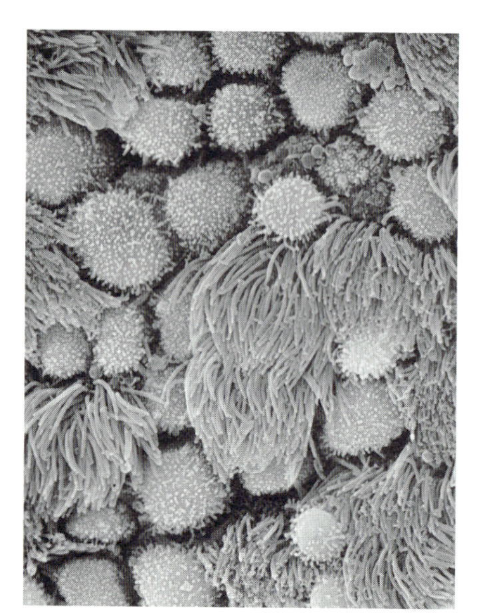

▲ Like certain cells in the sponge, the cells that line your trachea are covered with flagella-like hairs, called cilia, that beat in unison to move bacteria and particles away from your lungs.

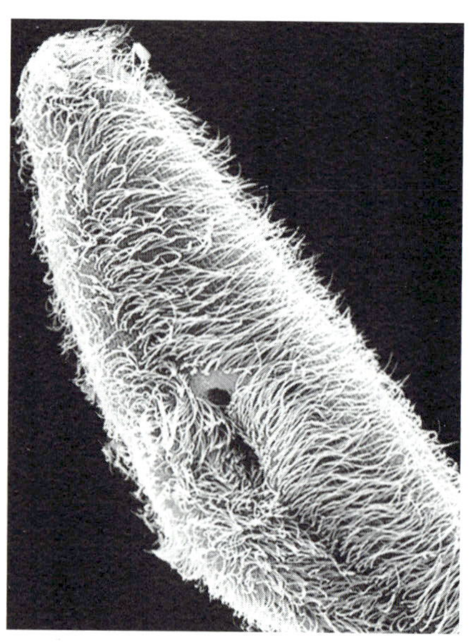

◀ This single-celled *Paramecium* is covered with hairlike cilia. Structured just like flagella, cilia move in unison to transport the *Paramecium* through the water.

2. Life Assembles Itself into Chains

◀ Concepts, p 4–5

"... *different* links ... are the alphabet of life."

More Information about Information

Vive la difference! Your hand is a useful tool for manipulating objects, but it can become an eloquent communication device if you use it to create the letters of American Sign Language (ASL). For the different hand shapes of ASL to qualify as information, they must be recognized by someone else, or they must be "read." As Helen Keller "read" the changing shapes of her tutor's hand, those shapes became information about the world. And, to be useful over time, information must be stored—placed in memory. For any system to have a memory, it must map differences in the world into coded sequences and keep these secure for later reading. Braille does this: it codes letters into distinct patterns of raised dots impressed on paper or other media.

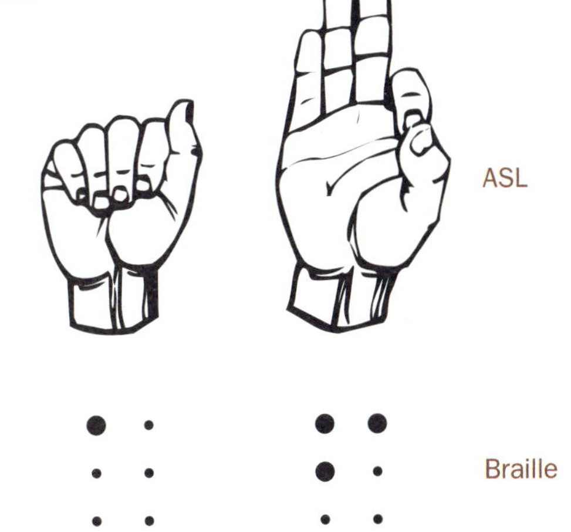

"... working chains, which carry out the business of living."

◀ Concepts, p 4–5

Reading Information

Living creatures have incredible capabilities for extracting information from—that is, reading differences in—the world. Monarch butterflies apparently navigate the 1,500 miles from Canada to a small area in Mexico by reading differences in the earth's magnetic field. Bats maneuver in darkness using echolocation: responding to differences in the echoes of the high-frequency sounds they emit. Trees "know" when to withdraw the nutrients from their leaves at the approach of winter, in part by recognizing differences in day length.

The ability of widely various creatures to read environmental information has a common source. Embedded in living cells are specialized chain molecules (proteins), which recognize and act upon tiny differences in their surroundings. These complexly coiled proteins act as information receptors and processors, picking up distinctive information from the environmental stream and reporting it to other working proteins for appropriate action.

3. Life Needs an Inside and an Outside

◀ Concepts, p 6

"... fat molecules have a water-liking head ..."

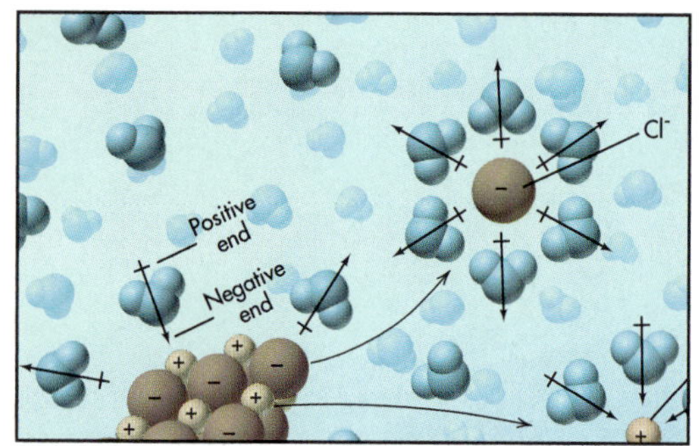

▲ Table salt crystals are composed of oppositely charged sodium and chloride atoms (called ions). In water, the negative chloride ion attracts the positively charged end of water molecules, and the positive sodium ion attracts the negatively charged end. These attractive forces are strong enough to pull the salt crystals apart. Notice how the dissolved salt components can now diffuse quickly through the water.

Water Organizes and Orients Other Molecules

Almost immeasurably tiny and opposing electric charges accumulate on the single oxygen and two hydrogen atoms of a water molecule. The attractions and repulsions set up by these electric charges create an environment for life. Water molecules cling to one another's oppositely charged ends, and so influence one another's orientation in space. They also attract or repel and therefore orient other kinds of charged molecules. This interplay between electrical attraction and repulsion sets the scene for the development of molecular containers (cells) that maintain an inner and an outer environment.

The molecules (called phospholipids) that make up most of a cell membrane have one charged end. The instant such molecules are in the presence of water, they orient their charged, oxygen-rich (phospho-) end toward the hydrogens of the water molecules. The fatty (lipid) tails of the molecules are not charged at all, and so tend to stay away from any water molecules. This hydrophilic ("water-liking") and hydrophobic ("water-shunning") molecular behavior provides a clue to how bilayered spheres might first have formed.

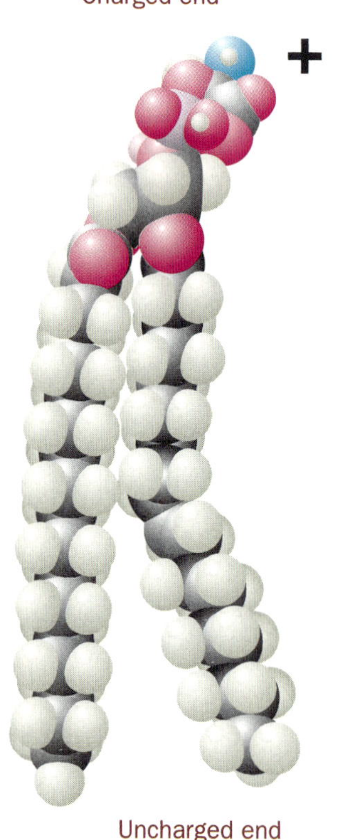

◀ A phospholipid molecule.

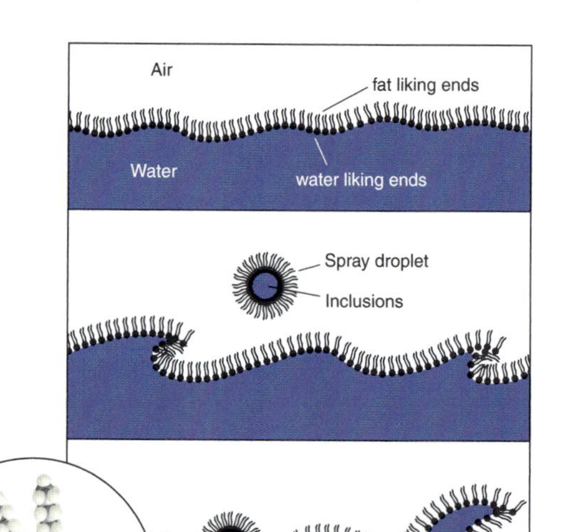

▶ When a layer of phospholipid molecules like this one on the water's surface is agitated by wind, spheres with water droplets inside can form, and when the sphere drops back to the surface, the fat-liking ends of the molecules on its outside join with those extending from the water surface to create bilayered spheres.

"Protein pumps . . . move molecules in and out."

◀Concepts, p 6

Reading the Landscape of a Cell Membrane

The phospholipid membrane that surrounds a cell is fluid, like the surface of a soap bubble, and has a lot of protein molecules embedded in it. The proteins that extend all the way through the membrane like the ones depicted here, act as gates, allowing selected materials to move in and out of the cell. Other proteins are attached to the outside of the bilayer and serve as identification or notification "badges". An approaching outside molecule must have a chemical shape and charge that locks on to the appropriate protein.

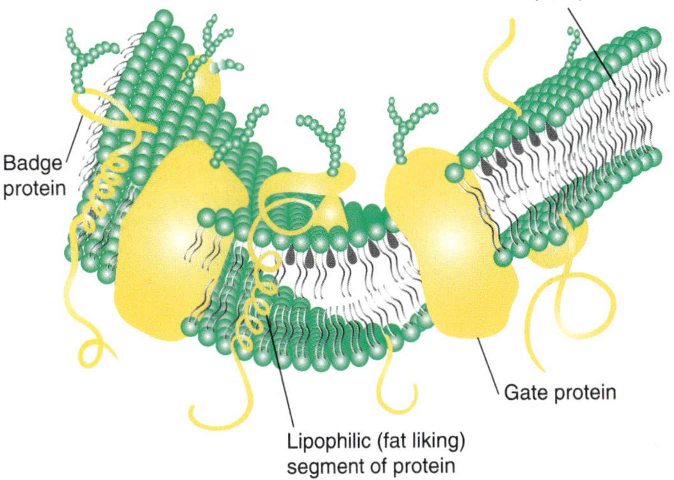

These badge proteins enable your body's cells to recognize each other, and notify the army (the white blood cells) to mobilize defenses against foreign invaders (such as bacteria). They also distinguish your blood type, which must be matched for a blood transfusion to be given safely. These proteins work to protect your cells from harmful chemicals that may enter your bloodstream in food or drugs; they do this by refusing to allow the harmful molecules through the membrane.

These same proteins can cause rejection of organ transplants because the transplanted cells have the wrong ID badges. Sometimes a certain type of badge will "overreact" to the presence of something like pollen grains or mildew spores, giving its unfortunate owner the runny nose and itchy eyes of a classic allergy attack.

Certain harmful invaders of the body, such as the HIV virus, use a sort of "counterfeit" badge to enter cells. The protein in charge allows this because a protein fragment on the virus locks onto it. Once inside the cell, the invader is provided with the raw materials it needs to replicate many times before it goes on its deadly way, as you saw on page 39.

?

These are the shapes taken by layers of certain dried protein molecules when they are heated slightly and mixed with water. Notice the double layers in (a) and the complex interior structures in (b). As it turns out, these membranes let certain substances through and repel others.

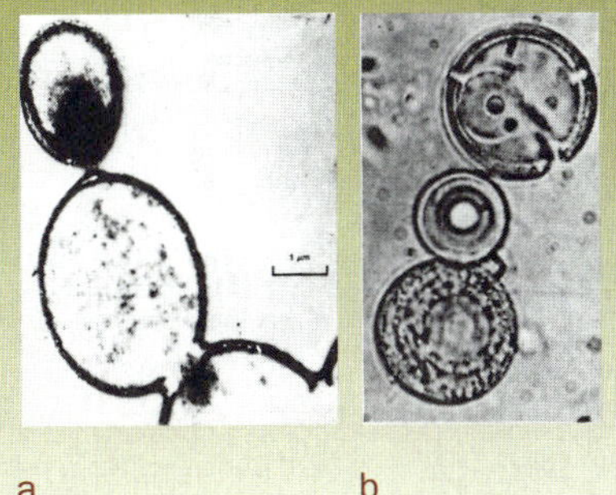

a b

Chemists studying the properties of phospholipid molecules and of fairly simple protein molecules found that when they were mixed with water, millions of them joined spontaneously to form small bubble-like spheres. When the mixture was shaken, the spheres broke up into even smaller spheres, always self-sealing. To scientists curious about the mystery of how cellular life might have started on our planet, this was an exciting discovery. Why?

Eureka! The Answer

The discovery provided a possible explanation of how the first cells might have arisen. A fundamental problem in understanding the origin of life is explaining how the vital chemicals might have been confined in a small enough space to promote continuous proximity of reactants. If natural substances like proteins or phospholipids spontaneously create self-sealing spaces, they might, in a primitive soup, have trapped these chemicals inside, allowing them to react and form molecules of amino acids and proteins.

"The atmosphere helps regulate
. . . protects . . . "

◀Concepts, p 6

▲ This NASA satellite photograph shows how incredibly thin the Earth's atmosphere is, compared to the huge volume of the planet itself. In this picture, the atmosphere appears to be just 2mm. deep.

Earth as Organism

Theorizing on a much grander scale than that of molecules and cells, some imaginative scientists have suggested that the earth itself can be viewed as a life form. In this view, called the Gaia hypothesis, the atmosphere, oceans, soils, and living organisms comprise a biosphere—a global self-regulating system that works to maintain its own internal balance (homeostasis) in much the same way as a cell or an organism does. Although this hypothesis is hotly contested in the scientific community, viewing the earth as a life form provides a useful model for thinking about living systems and their need for protective and containing membranes.

Terrestrial vegetation acts as a protective membrane for the land and its living contents. Vegetation gives off a great deal of the water vapor responsible for cloud formation and subsequent rainfall. When the membrane of trees and other plant life is removed from a region, water vapor is no longer given off and the surrounding land may become a desert. In coastal areas assured of plentiful rainfall, deforestation (removal of the vegetative membrane) leads to a result just as harmful; once the protective layer of plants and roots is disrupted, erosion is magnified with every rainstorm, and the nutrient layer of topsoil soon is washed away. Eventually little is able to grow or live there.

Earth itself is surrounded by a membrane that is both fragile and tough—the atmosphere. It admits light, vital to the existence of life on this planet, and emits excess infrared radiation (heat) produced by the activities of living things. The atmosphere protects us from the deadly cold and meteorites of space and from the Sun's harmful ultraviolet rays, moves and cleanses the air we breathe, and replenishes our fresh water supplies .

When the atmospheric membrane is perturbed (major volcanic eruptions, for instance, can launch particulates into the stratosphere), serious climatic imbalances can occur. After the violently explosive eruption of Krakatau in the South Pacific in 1883, volcanic dust in the stratosphere caused cooler temperatures and spectacular sunsets worldwide. The cooling effect was so great that 1884 was known as the "Year Without a Summer" in much of the Northern Hemisphere.

▲ A recent volcanic eruption seen from the Space Shuttle.

4. Life Uses a Few Themes to Generate Many Variations

◀ Concepts, p 8

"Life hangs on to what works."

How a Slime Mold Makes Its Living

We have already seen that anything living has to have an inside protected from the outside. And you know from the discussion of homology and analogy on pages viii and 34 that the shapes living things take over time are indirectly molded by specific survival needs and by the forces of the world outside. Most living things that are the wrong color or shape or size or have the wrong kind of teeth or breathing apparatus for their environment don't survive long enough to reproduce. Ones that are better adapted to their environment do reproduce and succeed.

The successful growth pattern in the yellow slime mold shown above right is an adaptation that allows it to absorb food from and exchange gases with the outside. It presents a very large surface to the world, and possesses a branching pattern of veins to move materials throughout its volume. The mold lives in a moist, dark, forest underlayer, so it doesn't need a thick shell or skin to protect it from its environment. In fact, in some sense, the forest underlayer can be thought of as the mold's "skin."

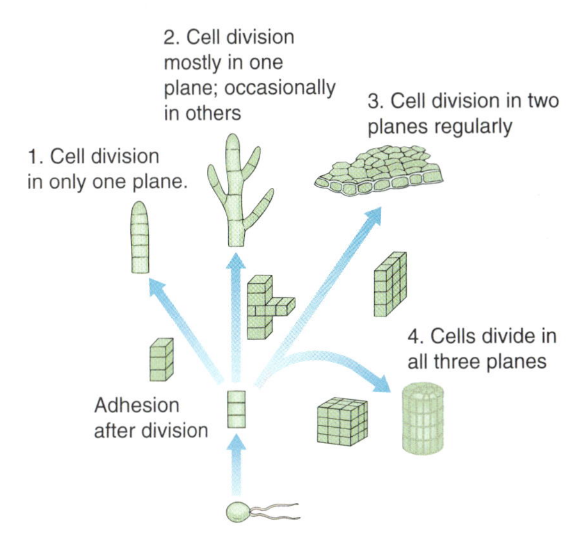

" . . . when cells divide and grow, they do so in a mere handful of ways."

Patterns of Multiplication

When a cell divides in two, which is how it reproduces, the two daughter cells can go their separate ways as unicellular organisms, or they can stick together and function as a multicellular organism. Dividing and adhering cells can occupy space in only four basic ways: (1) They can grow in one plane of space, say north-south, creating a single long chain of cells. (2) They can keep extending in that one direction, with occasional offshoots east or west. (3) They can grow consistently in two directions, making a thin sheet of connected cells. (4) Or they can grow in all three spatial planes, adding up and down to east-west and north-south, making chunks, cylinders, and spirals.

▼ Note the linear growth pattern of these strands of algae.

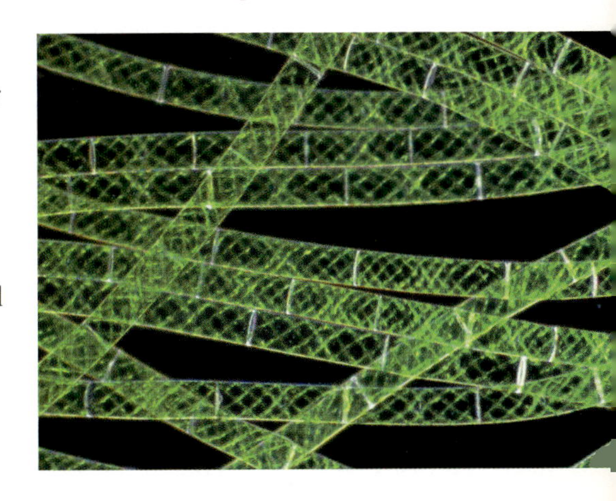

◀ Think of all the different patterns of cell division that created this dragonfly's shapes.

CONNECTIONS

The highly complex system of airways of a human lung has a pattern that is very similar to that of the simpler slime mold. A very different kind of organism, the seaweed *Fucus,* also has a similar pattern. **Explain the similarity of the patterns you see here in terms of the way each structure supplies the needs of the living organism. How are the outsides of these structure adapted to protecting them (or not) from their environment.**

Eureka! The Answer

The branching pattern of the airways in the lungs creates a large surface area for air to enter all parts of the lung. That same branching pattern exposes a large area of the seaweed to its fluid environment, allowing it to absorb nutrients and exchange gases. The "outside" of the lungs' air-ways is really their interior, and this is covered with cilia and mucus for protection. The seaweed, too, is covered with a kind of mucus, and has tough, leathery skin.

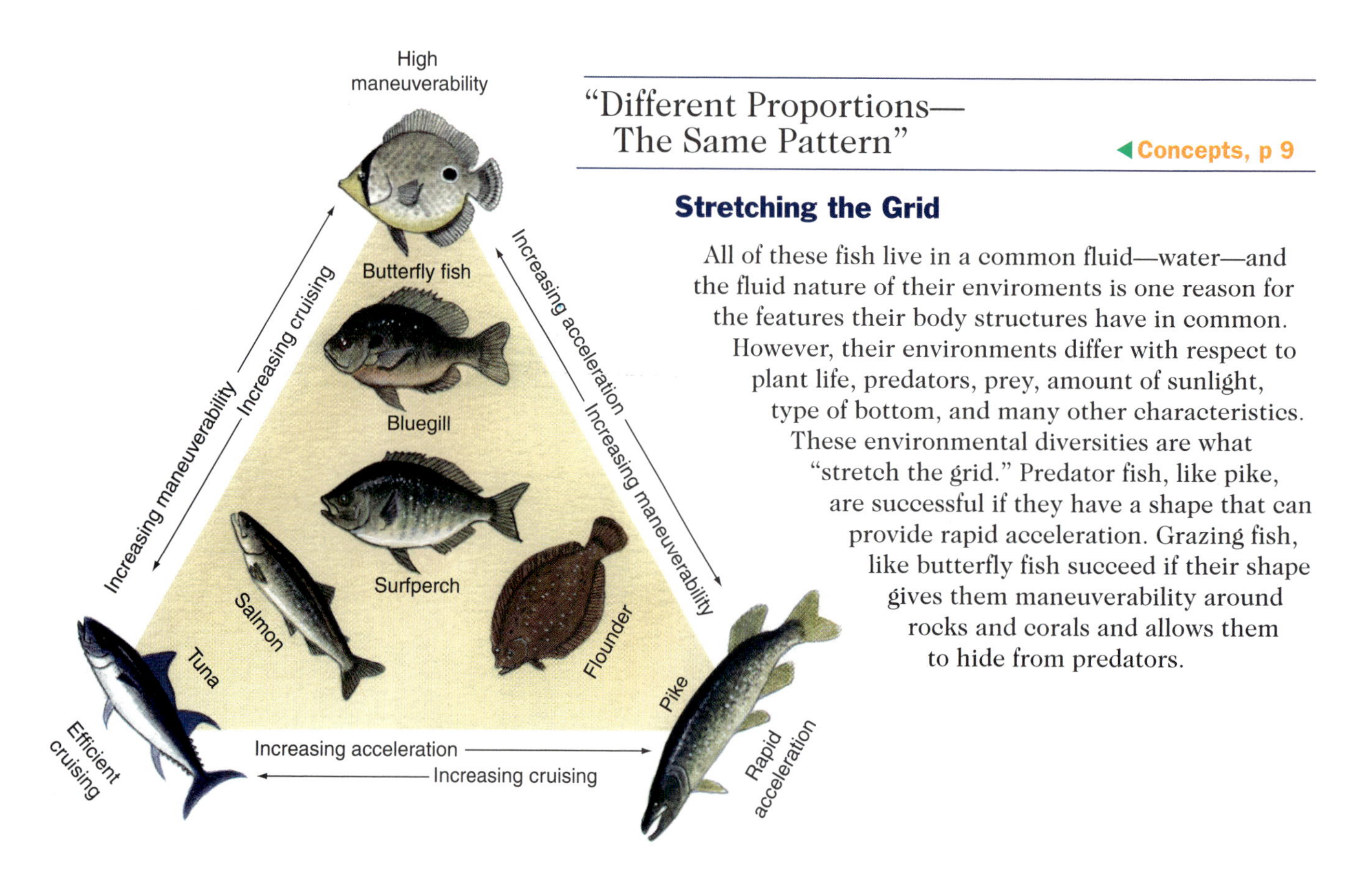

"Different Proportions—The Same Pattern"

◀Concepts, p 9

Stretching the Grid

All of these fish live in a common fluid—water—and the fluid nature of their enviroments is one reason for the features their body structures have in common. However, their environments differ with respect to plant life, predators, prey, amount of sunlight, type of bottom, and many other characteristics. These environmental diversities are what "stretch the grid." Predator fish, like pike, are successful if they have a shape that can provide rapid acceleration. Grazing fish, like butterfly fish succeed if their shape gives them maneuverability around rocks and corals and allows them to hide from predators.

CONNECTIONS

◀Concepts, p xx

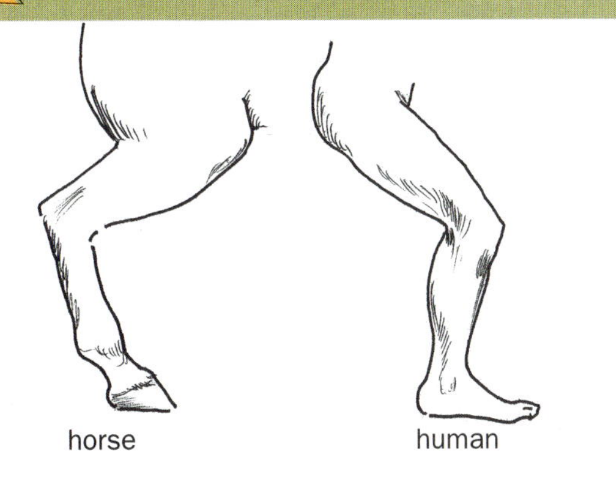

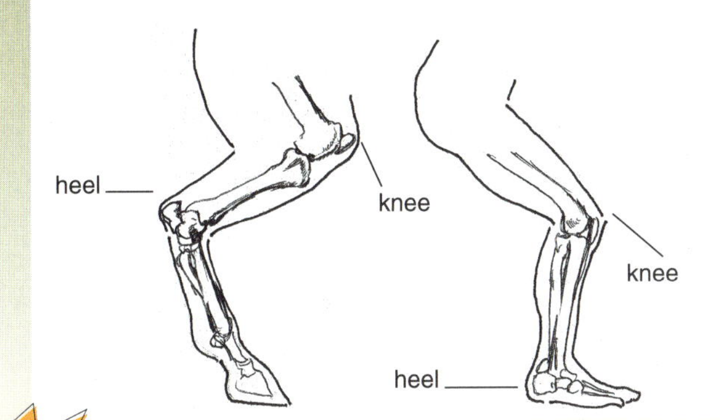

All mammals' skeletons are said to be fundamentally alike (homologous). These pictures show a horse's hind leg and a human leg. The two different legs appear to bend in the opposite direction. **Does this suggest a lack of homology, or is there some other explanation?**

Appearances are deceiving. A horse's hind leg bends forward at its knee, just as the human leg does. However, the proportions of the various leg bones differ—the horse actually stands on its toe.

If you were to take all of the beetles on pages 8 and 9 out of the widely diverse environments in which they actually live and place each, with a mate, on a rosebush with red blooms in the middle of a garden full of sharp-eyed songbirds, which of the beetles do you think would live to produce offspring? Which would be among the first to be eaten by the birds? Describe what the population of beetles might look like three years from now.

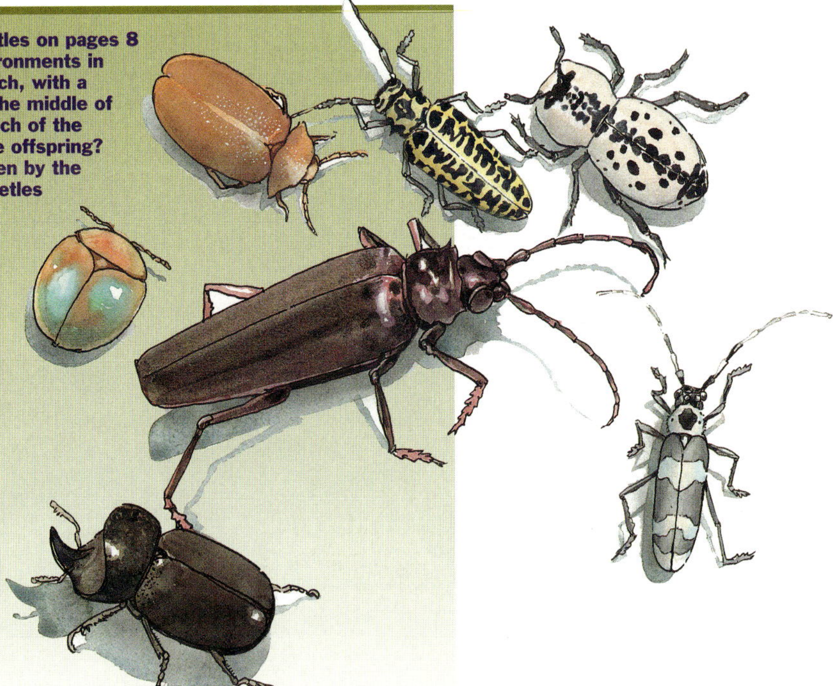

If the beetles stay on the leaves, the green ones should be more likely to survive, especially the smaller green one. The red ones might be okay if they stay on the flower petals. (Other qualities besides color might also affect, survival: how the beetles taste, whether they dwell on top of a leaf or on the underside, whether the predator birds are large or small, etc.)

5. Life Organizes with Information

◀Concepts, p 10

"... tiny sub-robots, each of which knew how to fabricate one stage ..."

Sweetly Splitting Sugar

The cell's workers—those sub-robots—are protein molecules (in this book, we often show them as the lumpy characters at right). They are able to perform relatively straightforward chemical tasks, like transforming a specific kind of molecule into a slightly altered version. They do this at incredible speeds—thousands of molecules processed per second—without being changed in the process. This kind of protein is called an enzyme—a biological catalyst, or a chemical reaction facilitator. Teams of enzymes work inside cells in a coordinated fashion to convert simple molecules, such as sugar, into the essential building blocks life uses to assemble its own substance: amino acids, nucleotides, fats, etc. Enzymes can also convert cholesterol to hormones such as estrogen and testosterone. They can, with the help of chemical energy, accomplish movement, as in the action of muscle or the transport of substances in and out of cells. They can, in sum, by their many coordinated interactions, maintain a human life!

A good example of a critically important life task accomplished by a team of enzyme workers is glycolysis (from the Greek "for sweet splitting"): the breakdown of the six-carbon molecules of glucose to two molecules of the three-carbon pyruvic acid (see top left next page). This process involves ten enzymes, each accomplishing tiny steps, and the end result is the net production of two energy-rich ATP (adenosine triphosphate, bottom of next page) molecules from one sugar molecule. ATP is the energy coinage of life, the molecule that is used by enzyme workers to drive all cellular activity.

Glycolysis is a way of getting energy from sugar when oxygen is not available. It was used by bacteria-like forms of life billions of years before

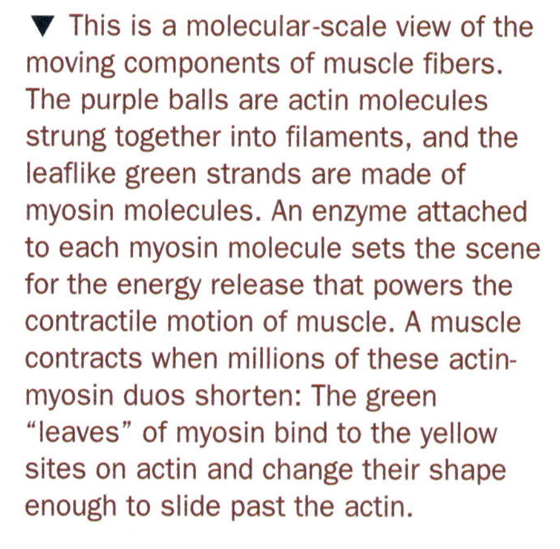

▼ This is a molecular-scale view of the moving components of muscle fibers. The purple balls are actin molecules strung together into filaments, and the leaflike green strands are made of myosin molecules. An enzyme attached to each myosin molecule sets the scene for the energy release that powers the contractile motion of muscle. A muscle contracts when millions of these actin-myosin duos shorten: The green "leaves" of myosin bind to the yellow sites on actin and change their shape enough to slide past the actin.

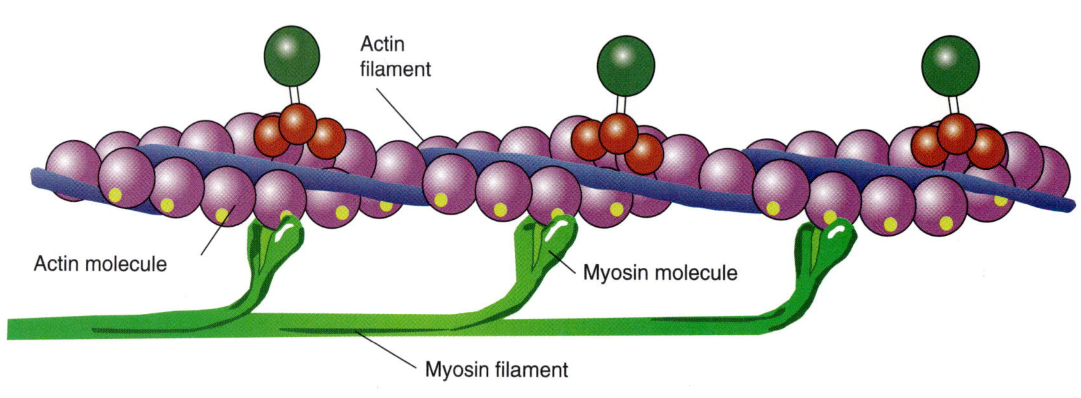

there was free oxygen in the earth's atmosphere. Organisms eventually evolved the ability to use oxygen to "burn" the waste product of glycolysis, pyruvic acid, to produce much larger amounts of ATP. But glycolysis proved to be so useful throughout evolution that it remains a central feature of almost all creatures alive today.

Glycolysis is but one example of many very simple parts working together in a coordinated fashion to create complexity. Seemingly impossible tasks are accomplished by doing one small chemical conversion at a time.

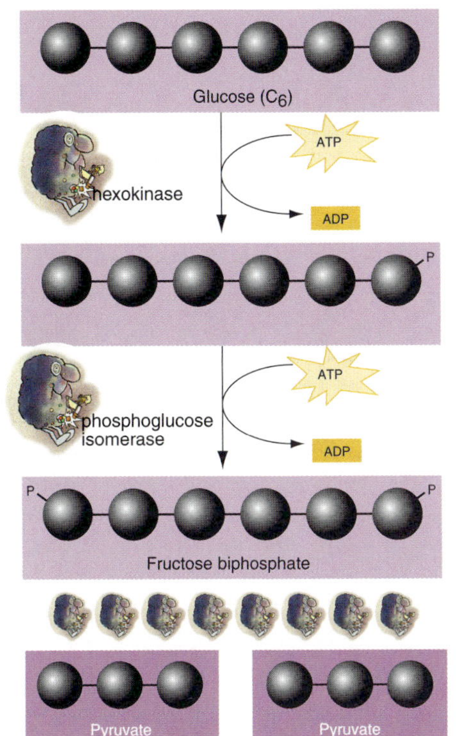

▲ This picture shows a simplified model of sugar—just its six carbon atoms. The enzyme hexokinase makes the first small change possible, the addition of a phosphate group (here shown as P). In the next tiny step, phosphoglucose isomerase, another enzyme, makes the next phosphate addition possible. Eight more enzymes orchestrate the remaining small chemical changes to glucose that turn it into two pyruvate molecules.

?

You're a robot designer and you've automated an extremely complex machine by designing little sub-robots to do various simple tasks. The problem is that these sub-robots are doing so many simultaneous tasks that they're getting in each other's way. If you could build a new feature into some of the sub-robots, they could orchestrate their various tasks and thereby reduce the chaos. **What new feature would you add? Can you come up with a cellular analogy for this robot design problem?**

Eureka! The Answer

The new feature would be a switch that turns a sub-robot on and off in response to a signal. The switch would be installed in at least one member of every production line or operating team so that it would function only when necessary. Similarly, body cells might reach a point where they were producing too much pyruvic acid from glucose. The solution to this problem might be another enzyme that reacts to that overload by interfering with the action of the enzyme hexokinase until the amount of pyruvic acid in the cell decreases.

Adenosine
Triphosphate

Adenosine
Triphosphate group
Sugar
Nitrogen base
NH_2
N
N
N
N
$^{-}O-P\sim O-P\sim O-P-O-CH_2$
O
C
H
H
C
H
H
C
C
OH
OH
High energy bonds

▶ Made of a sugar, three phosphate groups and a nitrogen base, ATP looks like the nucleotide molecules that make up DNA, though it has two extra phosphate groups attached to it. It is in the bonds that attach those phosphates that the energy is stored.

6. Life Encourages Variety by Reshuffling Information

◀Concepts, p 12

"... sexual reproduction, a more elaborate form of gene reshuffling."

Playing Games with Genes

As you saw on page 34, as changes in genes accumulate over time, the appearance of succeeding generations of a living organism can change dramatically. In nature, changes in genes arise from accidental alterations—mutations—and from the exchange of genetic information when organisms reproduce. With each alteration of the information that specifies their form and function, creatures become better or worse suited to their environment. The better suited survive to reproduce (to shuffle their genes again).

▲ ▶ By selecting only chickens with elaborate varieties of head plumage to breed with one another, breeders can quickly vary the appearance of successive generations of these showy chickens.

Gene reshuffling creates diversity, causing animals of the same species to come to vary greatly in appearance. Such variation within a species often occurs naturally, but it can be accelerated when genes are purposefully recombined through selective breeding. Throughout history, humans have taken advantage of the possibilities for variation allowed by gene reshuffling to control the characteristics of other animals. Dog breeders do this, for example. All domestic dogs, from Mexican Chihuahua to Great Dane, belong to the same species, though they look radically different. Cats, horses, sheep, and cows have also been manipulated genetically by humans to emphasize certain characteristics that make them more useful or decorative.

▼ Here you can see the progressively larger edible seeds of selectively bred corn plants. The wild form, called teosinte, is on the left of the picture.

Horticulturalists have been very selective in plant breeding, as well. U.S. grain crops, with their large seeds and huge yields, would be unrecognizable to ancient farmers. The flowers in almost every garden today are mostly selectively bred strains that didn't exist a century ago, and the same is true for most of the fruits and vegetables you eat.

?

The decorative chickens you see above are widely varying descendants of much plainer ancestral chickens. At Plimoth Plantation in Massachusetts, agricultural scientists are selectively breeding highly specialized modern chicken species to try to produce ones with the characteristics of their seventeenth-century ancestors—a kind of reverse selective breeding. **What characteristics would the scientists try to achieve, given the environmental challenges to the original Plimoth chickens ?**

Eureka! The Answer

The traits that would be selected for are likely to be muted colors, ability to thrive on sparse food, hardiness to cold and damp, quickness, and good vision.

7. Life Creates With Mistakes

◀ Concepts, p 15

"... copying errors... calling them mistakes oversimplifies."

Whether a Mistake is Good or Bad May Depend on Where you Are

Among humans, one genetic mistake shows up as sickle cell disease, a painful and debilitating hereditary condition. When depleted of oxygen (deoxygenated), normal human red blood cells retain their familiar round shape. In sickle cell disease, some of the deoxygenated red blood cells become elongated and curved in shape (like the tool called a sickle). When this happens, the sickle cells begin to clog blood vessels, and inflammation and tissue destruction occur. All of this damage is the result of a small change in the gene for the oxygen-carrying protein hemoglobin.

Interestingly, however, the sickle cell condition provides some protection against malaria, a blood parasite. Sickle cell disease is very common in equatorial Africa, where malaria is endemic. Scientists speculate that the disease, which might have been expected to remain uncommon because it decreases its victim's chances for survival and reproduction, is common precisely because it protects against malaria, a serious killer in Africa. (A person who is homozygous for sickle cell disease, meaning that the defective gene was inherited from both parents, usually dies at a young age; one who is heterozygous, or inherits the gene from only one parent, suffers a much milder form of the disease and thus lives long enough to enjoy its protective effects against malaria, and to pass the gene on to descendants.) This offers a good example of how a beneficial trait that evolved in one environment (tropical Africa) may prove detrimental in another environment (temperate America and Europe).

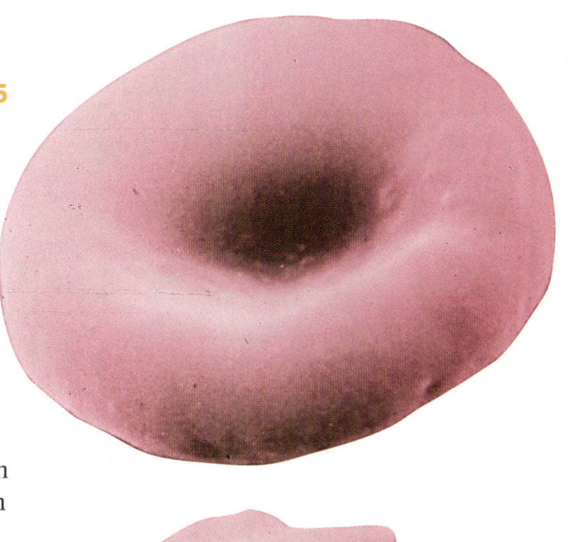

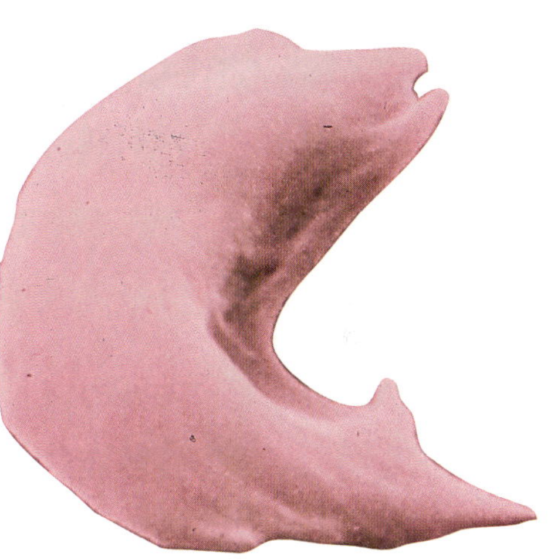

▲ These are scanning electron micrographs of a normal red blood cell (top) and of a cell with just one incorrect amino acid in its hemoglobin protein (see that protein on page 36).

"... every once in a while it shows up as an improvement..."

Mutations—Good or Bad?

Consider this paradox. The genes of all organisms have evolved to their present state through mutations—random changes in life's information chains (DNA). Mutations have gotten us here. At the same time, the vast majority of mutations are detrimental to the organism that inherits them. There are over 4000 known genetic diseases that are attributable to defects in a single gene. Huntington's disease, Down's syndrome, and sickle cell disease are examples of the harmful effects of mutated genes. Like a fine Swiss watch, perfected over centuries, an organism does not tolerate change easily. Modifications are likely to make it work less well, rather than better. Thus, it's not surprising that life has evolved mechanisms to

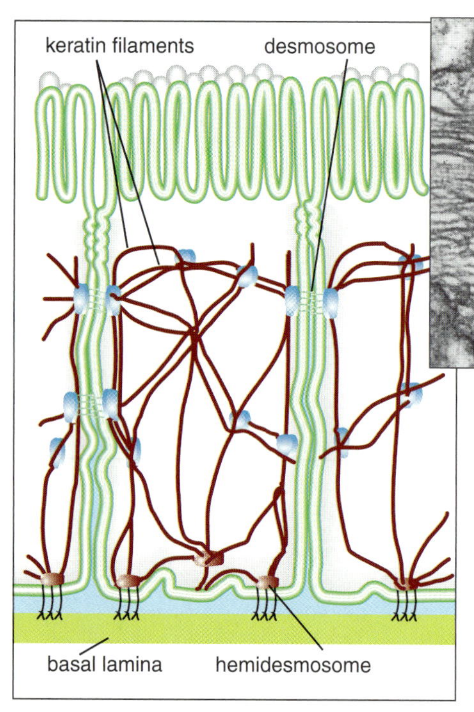

▲ Cells stick together by adherens proteins at places on their membranes called desmosomes.

avoid mutations. Cells have effective machinery for monitoring their DNA, finding errors and correcting them. So the mutation rate—the rate at which permanent, uncorrected changes accumulate in the genes of living creatures—is kept to a minimum. The mutation rate is estimated to be about 1 alteration per 1000 units of DNA every 200,000 years.

Mutations are generally caused by: (1) mistakes made by the DNA duplicating machinery when cells divide, (2) X-rays or UV and cosmic radiation impinging on DNA, or (3) certain toxic chemicals interacting with DNA. Viewed more broadly, a mutation could be considered to be any change in DNA, such as a mistake in duplication resulting in extra genes or the acquisition of genes from a virus. It is important to realize, however, that although a small change in a gene can affect a critical function of a protein, many changes in genes affect non crucial parts of a protein, and therefore have no noticeable effect.

It might seem ironic that genetic mutation, the source of disease or malformation, is also the source of evolutionary creativity. When the venerable and trustworthy Swiss watch underwent the substitution of a quartz crystal for its spring, it made a quantum leap forward in timekeeping. So it is with genes. A mutation that allowed cells to stick together opened the door for all multicellular organisms, both plants and animals. Such a change might have arisen when the gene for a receptor molecule in a single cell's membrane mutated and instead produced a protein that recognized and bound firmly to a receptor molecule on another cell. Proteins called adherens act in just this way.

Mutations affecting the gene for an ordinary housekeeping protein resulted in the transparent tissue that forms the lens of the eye. That protein, called lactate dehydrogenase, has been involved in cellular energy production for millenia. It is structurally identical to lens proteins called crystallins, which stack like lumber inside lens cells. The crucial genetic event in the past was probably a change that allowed large quantities of the housekeeping protein to be made. Somehow, cells packed with an excess of the transparent protein provided a simple organism of the past with the ability to detect and respond to light. An advantage that led, step by step, to the eye.

▲ The lens of the eye, as seen by a scanning electron micograph.

There's an important distinction to be made between the mutations that occur in the trillions of cells that make up the body and those that occur in the special cells that become sperm and egg cells. Mutations in body cells affect only the individual who sustains them. Cancer, for example, is aberrant growth of body cells caused by mutational damage to certain genes involved in control of cell growth. Cancer accounts for about a third of all human deaths. Mutations in sperm and egg cells, on the other hand, are passed on to another generation so a mutation rate substantially higher than the current one would compromise the future of the entire species.

8. Life Occurs in Water

◀ Concepts, p 16

"What is it about water that makes it so special?"

The Just-Rightness of Water

Water is by far the most plentiful chemical constituent of living creatures; it is the medium in which life came into being on our planet; and it is a pervasive part of the environment in which we now find ourselves. That molecules consisting of nothing more than an oxygen atom bonded to two hydrogen atoms could be so essential to sustaining life seems hard to believe, but water has some surprising properties that make it optimally suited to its role. 1) Its tendency, already mentioned, to become lighter (less dense) as it approaches freezing, and even lighter when it crystallizes as ice. 2) Its relatively high heat conductivity as a liquid but poorer conductivity as ice or snow. 3) Its thermal capacity—the fact that it takes a lot of energy to change its temperature. 4)Its high surface tension—cohesiveness or stick-togetherness. 5) Its relatively low viscosity—low resistance to flow and consequent ease with which substances diffuse through it. And finally 6) its capacity as an almost universal solvent.

This unusual collection of properties has a number of consequences for our environment: Water is preserved on the earth's surface (it doesn't fly off into space), and even in the coldest climates remains liquid under an insulating layer of ice and snow. Water has, for billions of years, crept into crevices of rocks (high surface tension) cracked them (when it expanded and froze), ground them up (in glaciers) dissolved its minerals and carried them to the sea in streams and rivers where they became essential constituents of living things.

Water contributes importantly to the general environment because of its resistance to temperature change: Vaporizing as the temperature rises (absorbing heat energy), condensing as the temperature falls (releasing heat energy). If water's density weren't close to that of the creatures floating in it, they would sink to the dark cold of the bottom or rise to the surface where they would be more readily exposed to the damaging effects of ultraviolet light.

Within and among living cells, water's optimum viscosity protects delicate structures from the sheering forces of shape change and motion—it acts as a kind of lubricant. Evaporation of water from the leaves of plants and trees constantly pulls water upward—water's surface tension makes this possible. Most of life's important molecules dissolve readily in water and diffuse through it rapidly to reach all parts of cells. (Oxygen, for example, diffuses through a cell in one hundredth of a second). In multicellular creatures, where a circulatory system is needed to move materials to all cells, and to conduct away waste and heat, water's low viscosity ensures that tiny diameter capillaries will conduct it and its dissolved chemicals to and from the remotest parts of the body. It is especially interesting that the viscosity of watery fluids containing cells (blood) drops as the pressure forcing it through a vessel rises, making the distribution of materials even easier.

It is hard to imagine a more perfect milieu for life. This is why scientists searching for clues to the existence of life elsewhere (as on the Moon, or Mars) keep a sharp eye out for evidence of past and present collections of water.

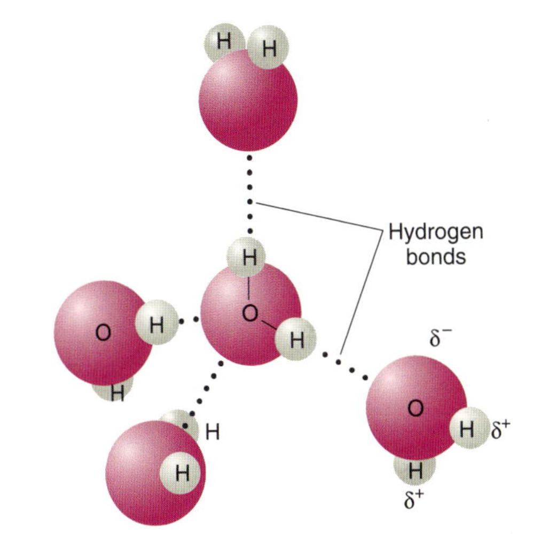

▲ ▼ The forces between the water molecules on the surface (surface tension) are strong enough to support the weight of this water strider.

Water's tendency to stick together means that water disrupted by dissolved salts on one side of a membrane will attract more water molecules to dilute the salt solution—a tendency called osmotic pressure.

One of these fish lives in the ocean; the other lives in fresh water. The labels describe each fish's water loss or gain and how it regulates the movement of water across its cell membranes. Which one of the fish must live in the ocean? What significance does the amount of urine production have?

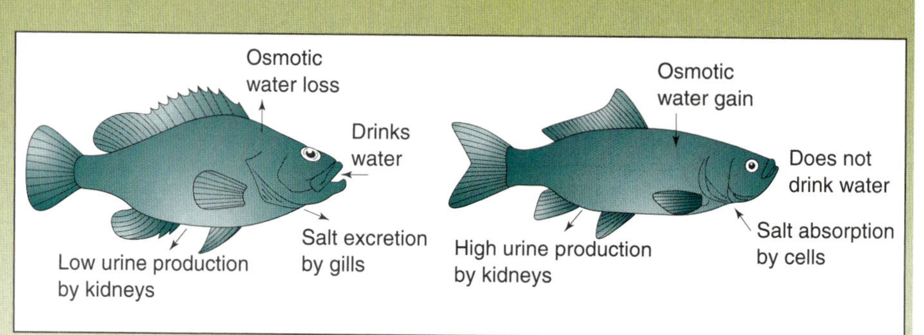

The fish on the right does not drink water, but still gains water through its cell membranes (osmotic water gain). It's exterior environment must be less salty than the interior environment of its cells; it is a fresh-water fish. High urine production is a way of concentrating salts in body cells to keep the internal environment from becoming as dilute as the exterior environment.

◀Concepts, p 18

9. Life Runs on Sugar

"Thus sugar percolates through life."

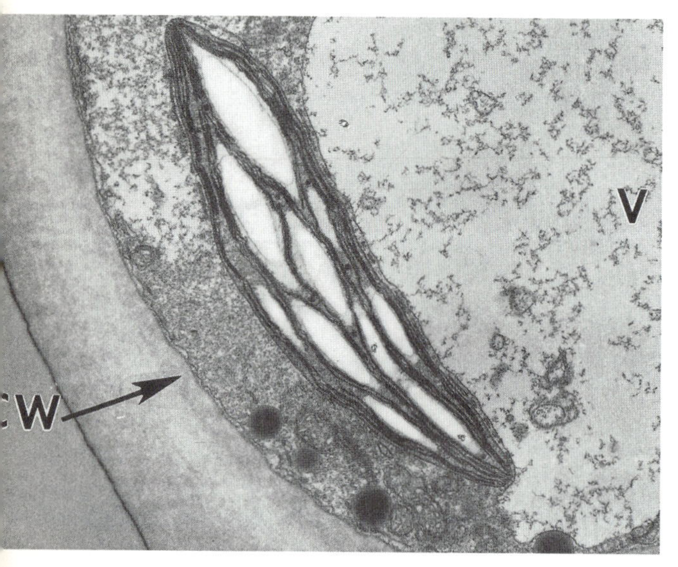

▲ In this photomicrograph of a plant cell, you can see the thick cell wall (CW), the cell membrane inside that, and a single chloroplast. The white shapes within the chloroplast are accumulations of long chains of starch molecules.

Trapping the Sun

Chlorophyll molecules in the chloroplasts of plant cells are trappers of sunlight. Just as water molecules have the right structure to absorb the energy of microwaves in an oven, chlorophyll's molecular structure absorbs the red and blue-violet wavelengths of sunlight (reflecting the green and yellow light to your eyes). A plant cell stores the solar energy trapped by its chlorophyll molecules in the chemical bonds of glucose molecules. Since a glucose molecule is quite small and stable when dissolved in water, it can move from cell to cell, wherever it is needed to supply energy. For long-term energy storage, glucose molecules bond together to form the long, bulky chain molecules of starch or cellulose.

Any living thing that can create its own food (glucose) is called an autotroph (self-eater). Autotrophs supply food for all other living things (heterotrophs, or other-eaters). Energy, stored in molecules' bonds, cycles through life and eventually re-emerges as heat.

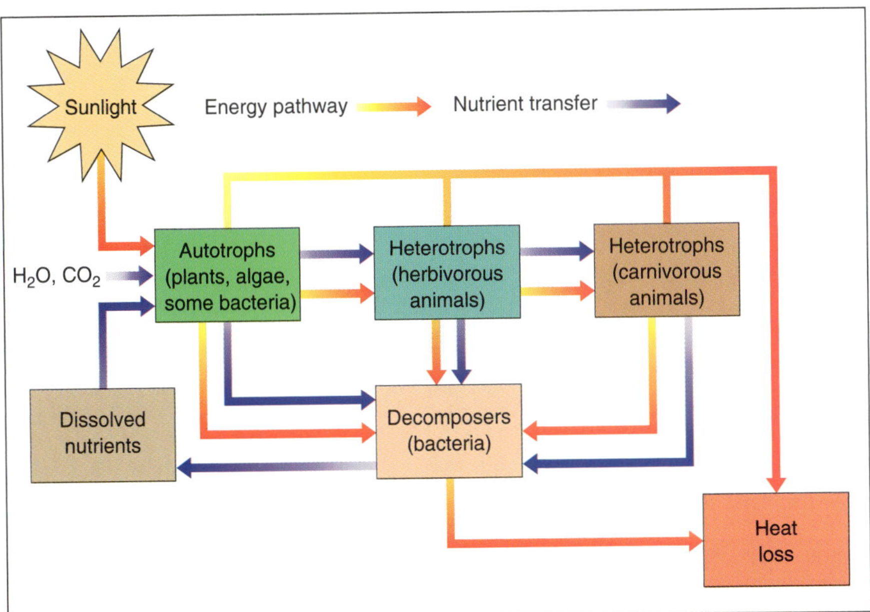

◀ The purple arrows show the cycle of food molecules through living things; the yellow-orange arrows indicate the fate of the sun's energy. Notice that the solar energy does not cycle. Once it has done its work of creating chemical energy within molecules, it is lost to the environment as heat.

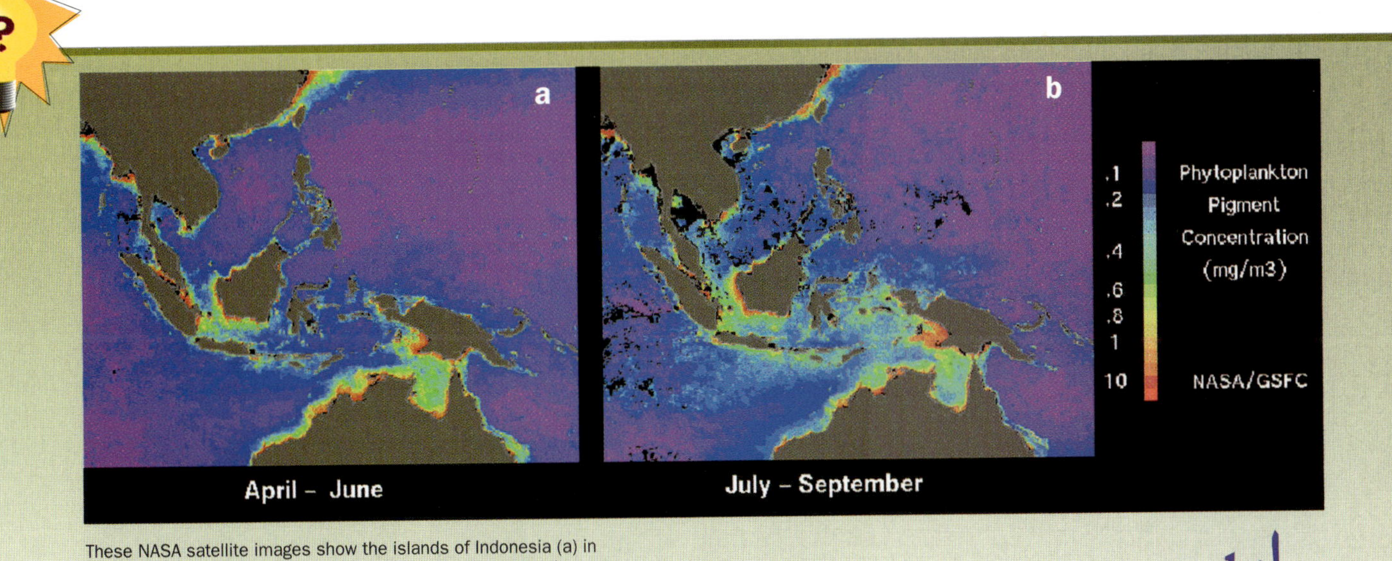

These NASA satellite images show the islands of Indonesia (a) in early spring (b) in summer. The red-orange through green colors indicate the presence of chlorophyll, contained in single-celled algae. **What do you think is the reason for the change that took place during the time between the two photographs. What could you predict about the relative amounts of oxygen and carbon dioxide above the ocean surface (a) and (b)?**

Eureka! The Answer

The red-to-green "bloom" is the result of a growth spurt of photosynthetic bacteria and plants in the ocean in summer when waters become warmer and enriched with nutrients that winter currents churn up from the depths. Since daytime photosynthesis uses carbon dioxide and produces oxygen, you could predict that there will be more oxygen and less carbon dioxide in the air above the ocean in (b).

▼ Here is another part of the feedback loop that suppresses infection. The green invading bacteria are engulfed by large white blood cells called macrophages ("big eaters"). The bacteria and the macrophages produce pyrogens and other proteins that cause immature white blood cells to mature into macrophages that can engulf more bacteria. This cycle continues as long as there are bacteria to trigger it.

10. Life Works in Cycles

"Loops tame uncontrolled events." ◀Concepts, p 20

Fever as a Feedback Loop

Fever is the body's biological response to viral and bacterial infection. Proteins called pyrogens, produced by the invading organisms and by the body's own white blood cell defense system, cause an area called the hypothalamus in the brain to "reset" the body's temperature higher. This reset signal constricts the blood vessels and causes shivers (which produce heat internally). The body's core temperature rises to the new set point, and the higher temperature either kills the invaders outright or stimulates the body's immune system to dispatch them.

Macrophages attack bacteria
Macrophage engulfs bacterium and displays parts on suface.
Bacteria
Proteins secreted
Specific immune response initiated
Body temperature rises
Monocyte
Monocytes triggered by proteins to mature into macrophages
Mature macrophage

When the invaders have been destroyed, pyrogen levels drop. The body responds by dilating the blood vessels and sweating profusely. Evaporation of water from the skin surface cools the body, bringing its temperature back to normal. This is a classic feedback loop—a self-correction par excellence.

?

A continuing feedback loop keeps the micro environment of this pond clean and nurturing for all kinds of organisms. Aquatic plants photosynthesize and provide food and oxygen for aquatic animals, which produce carbon dioxide as waste, and eventually die, providing further nutrients for microbes and plant life. When too many nutrients are pumped into such an environment from fertilizer used on crops or lawns, an uncontrolled event occurs; the loop is interrupted. In response to the nutrients, huge numbers of plants (algae, shown here on the lake's surface) grow, far more than the animal life can process. The plants die and fall to the bottom. In decomposing, the plants use up all of the water's dissolved oxygen. **What would be the overall effect of too little oxygen in the water? How could the interrupted loop be reestablished?**

With too little oxygen, fish and other aquatic animals suffocate. Their decomposition uses up even more oxygen. Limiting nutrients such as phosphorus and nitrogen that enhance plant growth and/or aerating the water to introduce more oxygen would help reestablish the pond's balanced ecology.

11. Life Recycles Everything It Uses

◀ Concepts, p 22

"... one organism's waste is another's food or building materials."

Ultimate Recycling

When things die, they don't go to waste. Death is a natural process for recycling life's raw materials. Life lets nothing go to waste. The carcass of a bison, left behind first by wolves and then by coyotes and crows continues to be digested and broken down by insects, bacteria, and fungi. When each of these organisms dies, it decomposes as well. Every living thing has other (often many) living entities that feed on it, and one organism's waste products are another organism's source of nutrients or shelter.

Cremation is simply combustion (burning), which turns the complex molecules that make up a living organism into simpler ones such as water and carbon dioxide gas. Thus the body's carbon, hydrogen, oxygen, nitrogen, calcium, and phosphorus atoms return to the atmosphere and the soil. Burial in an embalmed state, in a lined coffin, slows down the decomposition process for hundreds of years, and mummification can slow it for thousands. Eventually, though, every organism returns to the earth or atmosphere as a dispersal of atoms and molecules.

Every atom in your body is unimaginably old, dating back to the origin of the universe. However, during their last few billion years spent here on Earth, your atoms have cycled through a great many mineral and organic forms, over and over again. At one time or another, they may have been part of the atmosphere or the Amazon River, diatoms or dinosaurs, rocks or rabbits, trees or trilobites. This is ultimate recycling.

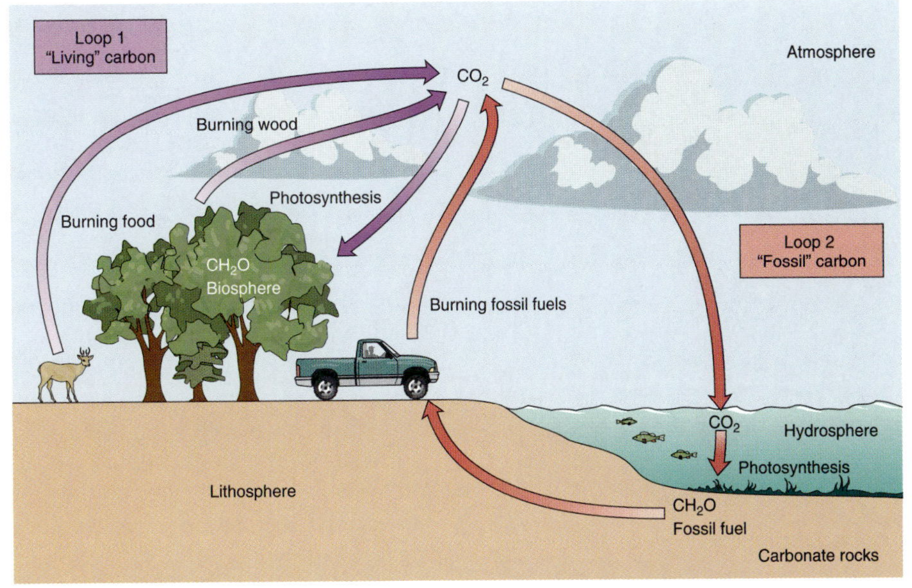

▲ Carbon atoms are the basic structural units of the molecules that compose the cellular structures of all living organisms. The purple arrows show how, by photosynthesis and respiration, carbon circulates through organisms and their environments. The carbon of long-dead organisms may be stored for millions of years in limestone (the remains of shellfish) or oil and coal deposits. Eventually, even that carbon is used as fertilizer, or burned and released to the atmosphere as carbon dioxide, once more becoming part of life's carbon loop.

▼ Here mushrooms and bacteria are hard at work decomposing the intricately complex molecular structures of these fallen trees.

◀Concepts, p 24

12. Life Maintains Itself by Turnover

"Cells that turn over in days or weeks..."

Hairs
Epidermis

▲ The bottom layer of these skin cells is fed by tiny blood vessels looping up from larger vessels, providing a constant supply of raw materials.

Parts renewal

Living cells' ongoing turnover requires a constant supply of raw materials and energy to assemble them. The energy, originally supplied by the sun and trapped in plants' molecular bonds by photosynthesis, enters your body in the form of food. That food provides most of the raw materials for synthesizing new proteins and other components of cells. Other raw materials come from the breakdown of these same complex molecules.

In some parts of the body, existing cells reproduce themselves by division. Your skin (or dermis) constantly replenishes itself, for instance, by shedding its surface layer (the epidermal cells). The cells at the base of the dermis (the basal cells) divide; as new cells are formed at the bottom and old ones are shed at the top, the cell layers advance toward the surface and eventual death. In fact, the surface of your entire body is covered with dead cells, forming a protective layer for the living cells beneath. What are they protecting you from? Ultraviolet radiation, chemicals, and abrasion. Living animal cells would die rather quickly in the presence of any of these.

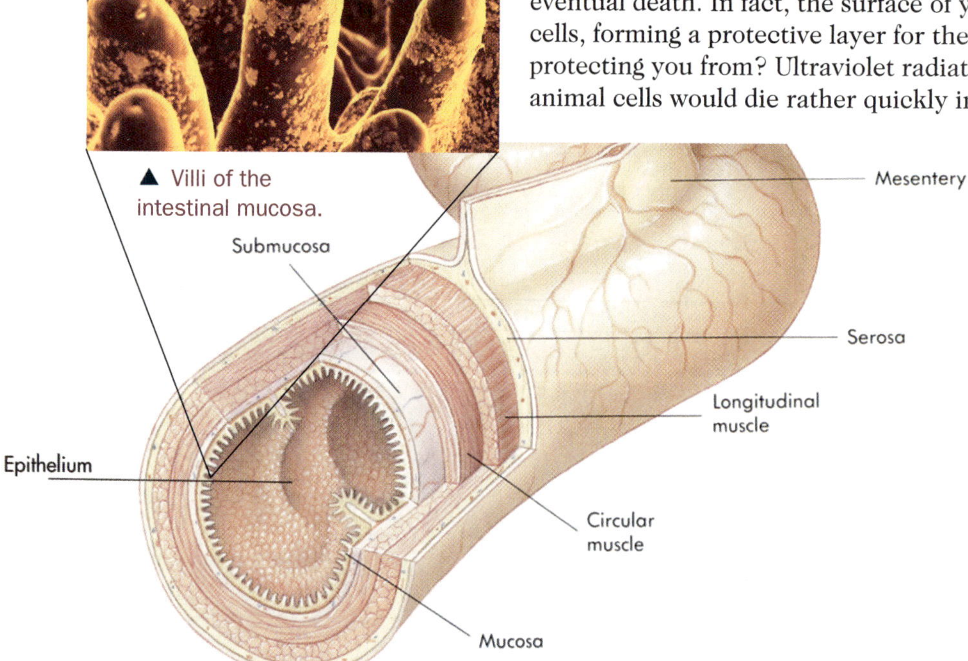

▲ Villi of the intestinal mucosa.

You have another dermal layer lining your gut (digestive tract) from mouth to anus. Your gut is effectively a tube that runs right through your body. Food goes in one end and is processed, and waste is expelled from the other end. The processing involves acids and enzymes that break down food substances into smaller molecules, and the dermal layer lining the gut protects your other cells from these digestive acids and enzymes. Each section of the gut is lined with specialized cells that variously protect, secrete chemicals, or absorb nutrients after food has been digested. The basal cells of the gut's dermal layer also divide continuously and the surface cells are scraped and sloughed off with the rest of the waste.

▲ In this section through the digestive tract (notice the many layers of cells that create an inside and an outside, and the finger-like projections (villi) that increase the surface area) starches and fats are broken down into smaller molecules, and those molecules diffuse through the fingerlike cells of the intestinal mucosa into the bloodstream. The blood carries these nutrient molecules throughout the body.

Blood cells, both red and white, originate in the marrow— ▶ the soft fatty tissue in the hollows of large bones. In the marrow, stem cells divide continuously, producing a constant supply of blood cells. A single drop of human blood typically contains 5 to 6 million red blood cells, and 5,000 to 10,000 white blood cells. Extra blood cells are stored in the spleen, and old ones are broken down in the liver — a constant turnover.

Provided with enough nourishment, your body can produce a sufficient supply of skin, gut, and blood cells. The information (DNA) in the nucleus of every skin cell is the prescription for the new cell. Too much sunlight can damage the DNA of skin cells, resulting in abnormal cells. When these abnormal cells divide, they often do so too fast, piling up into tumors. The liver, too, can regenerate itself if part is removed. Nerve cells also have some ability to regenerate. Most other cells in the body can't reproduce themselves. Instead, they maintain themselves by constantly breaking down and reassembling their proteins and other molecules.

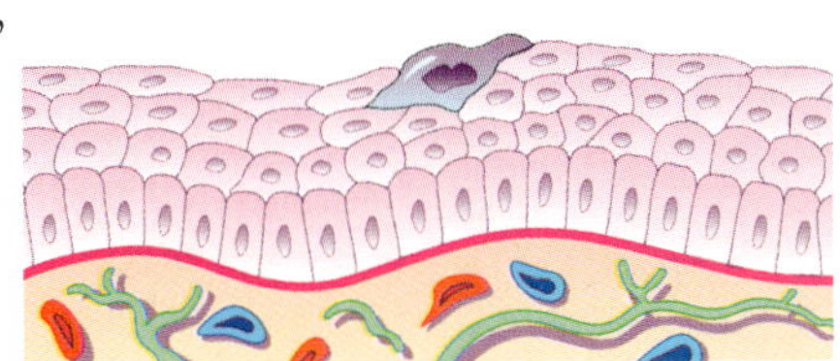
A single abnormal cell . . .

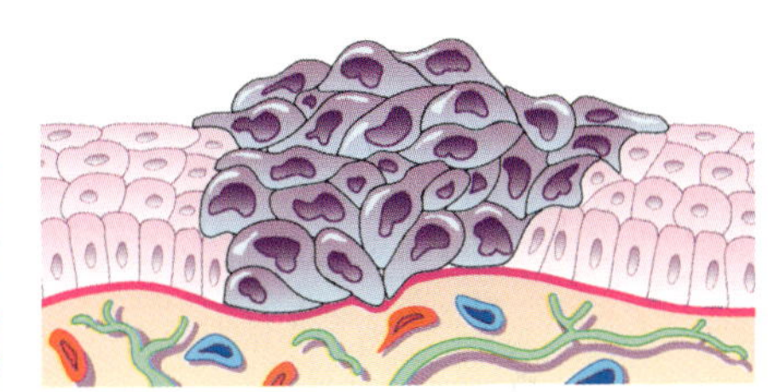
. . . becomes a tumor.

When a human liver is damaged in an accident and part of it has to be removed surgically, the remaining portion grows rapidly, producing a full-sized liver in a week or so. Experiments show that, normally, most of the liver's substance is being regularly broken down and rebuilt.(i.e., its big molecules are turning over inside its cells). Also, in the early stages of regeneration, the liver begins to increase in mass, without any increase in the rate of production of new cells. **How can cellular mass increase without a pickup in the rate of production?**

If there are increasing numbers of liver cells and the rate of production has not increased, it must be that fewer liver cells are being broken down.

13. Life Tends to Optimize Rather Than Maximize

◀ Concepts, p 26

"... a value in the middle range between too much and too little."

Less Is More

In times of crisis, the most specialized (maximized) organisms tend to become extinct; the most adaptable (least specialized) survive. Specialization always has a price: loss of adaptability. In stable times, maximization sometimes works; in changing times, optimization rules.

Over the course of life's evolution on earth, there have been periods of major global climate change and mass extinctions. Geologists have devised a time scale based on the sedimentary rock and fossil records. Two of the most famous documented mass extinctions came at the end of the Permian period, approximately 250 million years ago, and at the end of the Cretaceous period, 65 million years ago. It is estimated that more than 90 percent of marine animals and a large percentage of land animals became extinct at the end of the Permian. The Cretaceous extinction, most famous for the demise of the dinosaurs, saw the number of species decline by perhaps 50 percent.

In animals, many cells turn over (i.e., are born and die) but not most nerve cells (including those in the brain). **Speculate on why nerve cells rarely turn over.**

One plausible theory is that nerve cells' function - including memory - depend on patterns of connections with other nerve cells. These connections develop over time. Cell death and replacement could break existing connections, thereby disrupting memory and other functions.

?

With all hair removed, the bodies of a gibbon and a human are remarkably similar. Humans have adapted successfully to far more environments than have gibbons. **How is the gibbon maximized compared to the human? How might the human's optimal body structure explain its success in adapting?**

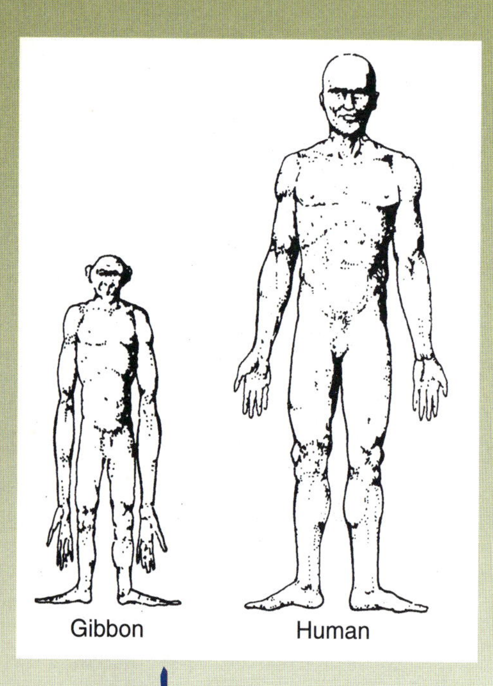

Eureka! The Answer

The gibbon's arms, hands and legs are optimized only for living in and moving through treetops—outside a forest these traits would be maximized, and a gibbon would be largely helpless and easy prey for fast ground-based predators. Also, the use of the hands for locomotion interferes with their usefulness as a tool for manipulation of objects in the environment. The human, on the other hand, while not specialized for a single environment, can climb trees if necessary, run across open land, climb mountains, and use tools to modify his environment.

What survived? The least specialized, most adaptable organisms. In other words, the optimized plants and animals capable of living in altered environments, adapting to changes in climate, air or water composition, and diets. Following a mass extinction, adaptive radiation (the evolution of many different species from a few ancestors) occurs on a large scale as life rushes to fill the vacated niches of vanished species.

Some of the most spectacular fossils are those of maximized species, such as the largest dinosaurs, the saber-toothed tiger, the woolly mammoth, and the Irish Elk pictured on p. 27. But maximization is not confined to large animals. Certain plant species (many orchids, for example) depend entirely on a single species of insect for pollination. The two, plant and insect, are said to have coevolved. The disappearance of the one is likely to lead to the extinction of the other.

▲ Only certain insects can fit this orchid's distinctive shape, and the orchid's pollen is deposited on very specific parts of such an insect. When the insect visits the next orchid, pollen is rubbed directly onto the plant's reproductive part, the stigma.

Certain aphids demonstrate a sort of optimization. One species (pea aphids) produces individuals of two colors, red and green. Both colors exist together, feeding on the same plants (peas and other legumes). The pea aphids have two main predators, ladybugs and parasitic wasps. The ladybugs primarily eat the red aphids, presumably because they're easier to see. The wasps more often lay eggs in the green aphids, perhaps because eggs laid in the red individuals get eaten before maturity. The dual coloration appears to be an optimization strategy, permitting more individuals to survive in the presence of either predator.

14. Life Is Opportunistic

"... life flourishes even in the world's harshest places"

◀Concepts, p 28

Bacterial chloroplasts

In the absolute darkness of the ocean bottom, no photosynthesis can occur. Even so, dense communities of gigantic tube worms, mussels, and

clams cluster around towering hot water vents on the ocean floor. These communities are something of a mystery. There are absolutely no plants here. What do these creatures use for food? Why are they found only near the hot water vents?

It seems that the answer lies in the molecules dissolved in the hot sea water. As sea water seeps into the ocean floor in places where hot magma from the earth's interior is close to the surface, it dissolves subterranean minerals such as iron, calcium, sulfur and copper. These minerals then rise through the vents in the ocean floor with the heated water, providing food and energy (in the form of the bonds in molecules of hydrogen sulfide gas) for enormous communities of bacteria that live on the inner surfaces of the vents. Just as the chloroplasts in plants use solar energy to turn carbon dioxide and hydrogen into sugar, these bacteria use the chemical energy in hydrogen sulfide molecules to make sugar. The tube worms, mussels, and clams filter these bacteria out of the sea water, ingest them, and use their sugars as nourishment.

▲ A colony of tube worms.

15. Life Competes Within a Cooperative Framework

"Nice guys last longer."

◀ Concepts, p 30

Shrimp Being Selfish

In the June 6, 1997 issue of **Nature**, J. Emmett Duffy reported his observations of several different colonies of Caribbean snapping shrimp. He found that all of the shrimp in a single colony (they live inside sponges) are closely related—descendants of a single mother, or "queen," and a single father. The shrimp are fierce defenders of their colonies and will chase off or kill any intruder's from other colonies. Duffy set up some small experimental colonies in his lab and discovered that the shrimp welcomed former members of their own colony, even when space and food were in short supply. This welcoming behavior might seem to be foolish at first, but it is a beautiful example of combined self-interest and cooperation. Purely individualistic self-interest might dictate that any intruder would use up precious resources and should be killed or chased off. The longer-term interest of the colony, though, would consider any shrimp with the same DNA just as valuable for carrying on the life of the colony. Thus, the welcoming behavior is in the interest of the colony's survival.

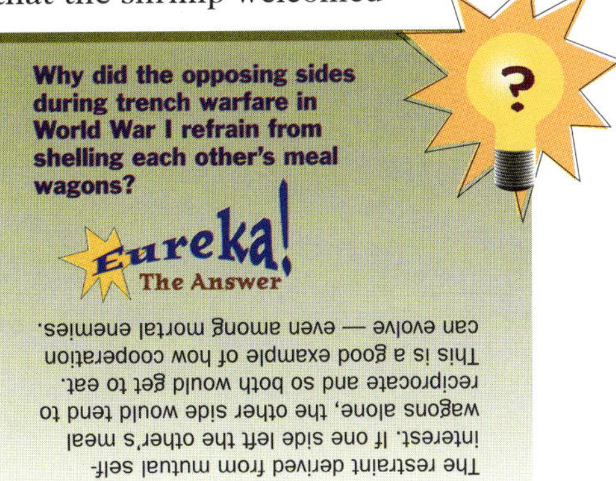

▲ The Earth and its moon, as seen by the Galileo spacecraft, show the vast difference that bacterial life has made to the surface of our planet.

▲ This photo of Mars' surface was taken in 1997 during the Pathfinder landing. This mission confirmed that Mars had once been covered by large amounts of liquid water, but so far there is no undisputed evidence that there has been life on the planet.

16. Life Is Interconnected and Interdependent

"... a multilevel, integrated system." ◂Concepts, p 32

Small Creatures, Big Effects

New evidence points to the astonishing role that life, particularly microscopic life, has played in establishing and maintaining the earth's atmosphere, temperature and climate. Life has acted both as a stabilizing force, dampening the effects of solar fluctuations and volcanic activity, and as a creative force, setting the stage for new and more complex organisms.

For example, estimates of the sun's energy output in life's earliest phase suggest that the earth would have frozen over but for the small amounts of ammonia released by primitive bacteria. These were simple creatures that thrived in a low-oxygen environment. Later, photosynthetic bacteria and other organic matter secreted enough oxygen into the atmosphere to create an environment in which aerobic (oxygen-breathing) creatures, including the first animals, evolved.

Since the earth's beginning, the sun has been getting hotter. At the same time, volcanoes have been adding carbon dioxide to the atmosphere, trapping heat that would otherwise escape into space (the well-known greenhouse effect). Although the earth's temperature would be expected to rise steadily, this has not happened—at least until the widespread burning of fossil fuels in this century. Instead, average temperature has remained relatively constant, primarily because, over eons, microbes have steadily and dramatically reduced the amount of carbon dioxide in the air to its current level of 0.03 percent. Tiny plants, as they grow and die, "nibble away" rocks and trap carbon dioxide in the soil. There, it dissolves in water and washes eventually into the sea where it is used by marine life to build shells. Ocean algae also trap large quantities of carbon dioxide, which finds its way into marine shells (as calcium carbonate) and ends up on the ocean floor, ultimately becoming limestone and chalk.

And what about water? Planets such as Mars and Venus have lost all of their surface liquid water, probably because certain elements such as iron bonded with the water's oxygen to form oxides. The leftover hydrogen was too light to be held by the planet's gravity and was lost to space. On earth, microbes prevented this loss of hydrogen by taking volcanic hydrogen sulfide, extracting the hydrogen for energy, and excreting the sulfur as pellets. Also, photosynthetic organisms kept producing new water—enough for the earth to retain its oceans, and thus, provide an environment for the further evolution of life.

A recent and plausible theory proposes that microorganisms are responsible for cloud formation over the oceans. Cloud formation is aided by the presence of tiny particles. Marine algae emit vast quantities of sulfide particles, which serve as "seeds" for cloud condensation. The creation of clouds over oceans covering two-thirds of the earth's surface significantly affects the global climate.

So microscopic forms of life have worldwide physical and chemical effects—they create much of their own environment.